TACKLING FARM WASTE

DEDICATION

THIS BOOK is dedicated to John and Winifred Grundey, my parents, for a lifetime's encouragement and to my wife Beryl for making life the pleasure it is.

TACKLING FARM WASTE

KEVIN GRUNDEY

FARMING PRESS LIMITED
WHARFEDALE ROAD, IPSWICH, SUFFOLK

First published 1980
ISBN 0 85236 103 3 ✓

© FARMING PRESS LTD 1980

All rights reserved. No part of this publication may be reproduced, stored in a retrieval system, or transmitted, in any form or by any means, electronic, mechanical, photocopying, recording or otherwise, without the prior permission of Farming Press Limited.

Set in ten on eleven point Times and printed in Great Britain on Longbow Cartridge paper by Page Bros (Norwich) Ltd for Farming Press Limited.

CONTENTS

CHAPTER	PAGE
FOREWORD	11

By David Richardson
White Rails Farm, Great Melton, Norwich

AUTHOR'S NOTE 13

ACKNOWLEDGMENTS 14

1 INTRODUCTION: BASIC PROBLEMS WITH FARM WASTE 15

Avoiding water pollution—danger of unduly strict controls—manure production.

2 METRICATION AND GLOSSARY OF TERMS 22

Metrication—decimal multiples and sub-multiples.

3 LEGISLATION 30

Fifteen Acts implemented or partly implemented.

4 MANURES AND SLURRIES 41

Volumes of animal waste produced—materials added to manure. Bedding materials: poultry; pigs. Water: cleaning water; rain water; Meteorological Office data; underground seepage. Farmyard manure—plant nutrients in manures and slurries—nutrient loss during storage—FYM losses—seepage—losses from slurry—availability of nutrients in manures—financial value of manures.

5   STORAGE   66

Storage of FYM—storage of slurry—advantages of slurry and drawbacks. Storage systems: lagoons; site; construction. Compounds: siting; sizing; earth-banked compound with earth floor; emptying compounds. Floor; earth-banked compound with concrete floor; concrete bottom with sides designed to allow liquid to escape; straw-bale compounds. Additional facilities for use with compounds: loading ramp; strainer box; fencing. Concrete tank structures: concrete block below-ground tanks for pigs; cattle; poultry. Tanks constructed of concrete sprayed in situ—concrete panel stores—stores made of concrete pipes. Above-ground steel slurry stores; siting; construction; accessories for mixing—hydraulic systems—pneumatic systems—mechanical systems; management of above-ground stores.

6   SLURRY AND MANURE HANDLING   104

Handling in and around buildings: mechanically unassisted handling—overflow slurry channels—channel sizes—'V'-shaped slurry channels—hydraulic flushing; mechanically assisted handling—handling solid material—automatic scrapers. Handling slurry with pumps; augers; elevators; and pneumatics. Handling to the field: tankers—types and sizes—wheels and brakes—discharge pattern—power requirements—work rate—costs. Soil injection of slurry; equipment—work rate. Organic irrigation of slurry. Work of FYM loaders—transporting and spreading FYM—side delivery rotary spreader—rear delivery spreader. Calculating application rates of manures. General-purpose trailers.

7   SEPARATION   149

Products of separation—separated fibre—separated liquid—fertiliser value of separated slurry. Separator machines; centrifuge; vibratory screen; roller press machine; belt press; combined gravity screen and compression machine. Siting the separator—feeding the separator—dilution—disadvantages and advantages of separation.

8   AEROBIC AND ANAEROBIC TREATMENT OF MANURE   165

Aerobic treatment—heat recovery—aerobic and anaerobic treatment of solid manure—anaerobic digestion of slurry—pH level—temperature—solids retention time—gas quality—digester—utilisation of biogas energy—safety—plant operation—costs.

9 UTILISATION OF MANURES AND WASTES 182

Manures as a fertiliser—manures for re-feeding—manure drying—hazards. Ensiling poultry litter techniques—feeding ensiled poultry litter. Pig manure re-feeding.

10 OTHER FARM WASTES 197

Estimating yard and washwater volumes: partial treatment; storing yard effluent for land disposal. Complete treatment. Barriered ditch system: siting; construction; maintenance; safety. Dead livestock; drums and containers; metal scrap; milk and milk products; Oil. Silage effluent—quantity produced—collection and storage—poisonous gases—utilisation as a fertiliser and as a feed. Spray and other surplus chemicals: sheep dip. Straw burning. Vegetable washing water—minimising the problem—systems for disposal—reduction of suspended solids—achieving dischargeable standards. Non-farm wastes: brewers' grains; swill.

11 AGRICULTURAL ODOUR 228

Smell: what is it? Sources and control of odour—about buildings and from field operations. Aeration—anaerobic control—chemical control—field equipment for control. On receiving an odour complaint.

INDEX 245

ILLUSTRATIONS

PLATES	PAGE
1. Lagoon storage	62
2. Above-ground manure storage system	62
3. Wall bracing and seepage collection gutter (see Plate 2)	63
4. Concrete panel store	63
5. Careful siting of large slurry store	64
6. Slurry agitation by recirculating slurry under pressure	64
7. Slurry store agitation by air bubbling system	65
8. Slurry agitation showing position of compressor and piping	65
9. Four-wheel drive forklift truck at work	113
10. Cleaning out litter from broiler shed	113
11. Motorised assistance with mucking out	114
12. Automatic mechanical scraping	114
13. Pto-operated slurry pump	115
14. Close-up of auger cutter of slurry pump	115
15. Slurry tanker with quick-action hose attachment	142
16. Slurry tanker at work on autumn stubble	142
17. Close-up of tines of rear-mounted slurry injector	143
18. Slurry injector	143
19. Slurry injection frame and slurry tanker in parallel operation	144
20. Travelling reel type of irrigator	145
21. Irrigation gun mounted on wheeled sledge	145
22. Side delivery rotary spreader at work	146
23. Rear delivery muckspreader	146
24. Typical slurry separator installation	159
25. Main elements in slurry handling and separator installation	160
26. Type of separator that produces driest fibre	161
27. Low-volume travelling irrigator for separated slurry liquid	161
28. Showing width of boom and evenness of spreading (see Plate 27)	162
29. Simple dribble bar applicator for separated liquid	162
30. Floating surface irrigator	176
31. Aerator (see Plate 30) in action	176
32. Large surface irrigator	177
33. Floating sub-surface down-draught aerator	177
34. 14 kW aerator (Plate 33) at work in pig slurry	178
35. 227 m^3 anaerobic digester treating pig slurry	178
36. Inside of generator hut (Plate 35)	179
37. Below-ground anaerobic digester treating cow manure	179
38. Anaerobic digester built by Dutch farmer	191
39. The Fiat TOTEM unit	191

40.	TOTEM unit showing generator and a waste heat exchanger	192
41.	Bank of TOTEM units running on biogas	192
42.	Drying poultry manure beneath laying cages	193
43.	Another way of drying poultry manure under cages	193
44.	Dribble bar attachment for Farrow slurry tanker	194
45.	Weeks slurry curtain attachment for tankers	194

DIAGRAMS

Fig. 1.	Dairy cow population 1970–78, England & Wales	16
Fig. 2.	Pig population 1970–78, England & Wales	17
Fig. 3.	Laying hen population 1970–78, England & Wales	17
Fig. 4.	Farm labour force 1970–78, England & Wales	18
Fig. 5.	Earth bank compound	81
Fig. 6.	Straw bale compound	85
Fig. 7.	Loading ramp for slurry compound	86
Fig. 8.	Welded mesh slurry strainer box	88
Fig. 9.	Sluice gates dividing long slurry channels	93
Fig. 10.	Beef building on slats	94
Fig. 11.	Battery cages and deep pit poultry manure storage	96
Fig. 12.	Slurry graph DM, consistency and handling equipment	105
Fig. 13.	Nomograph of water needed to change slurry dry matter	106
Fig. 14.	Overflow slurry channel	106
Fig. 15.	Overflow slurry channel—detail of cleaning out drain	107
Fig. 16.	Overflow slurry channels—alternative arrangements	108
Fig. 17.	Pneumatic 'letter box' slurry pump	124
Fig. 18.	Temperature of separated solids when composted	151
Fig. 19.	Input slurry DM % and separated fibre DM %	152
Fig. 20.	DM % of resulting liquid from same slurry (see fig. 20)	152
Fig. 21.	Roller press slurry separator	156
Fig. 22.	Rainfall, catchment area and resulting total volume of rainwater	198
Fig. 23.	Settlement tank (below ground)	200
Fig. 24.	Profile of barriered ditch	203
Fig. 25.	Barriered ditch: timber sleeper barrier	204
Fig. 26.	Amount of effluent produced in silage-making	211
Fig. 27.	Rate of silage effluent production	212
Fig. 28.	Farm-built sediment tank for vegetable washing water	223
Fig. 29.	IMAG-designed biological air washer for piggeries	233
Fig. 30.	Axial-flow aerator/mixer	235
Fig. 31.	Upward throw floating aerator	236
Fig. 32.	Down-draught floating aerator	236

FOREWORD
by DAVID RICHARDSON

I FIRST got to know Kevin Grundey when he was a general machinery adviser for what was then NAAS in Norfolk. Since those days of the 1960s he has specialised in all aspects of farm waste and is now a leading authority on the subject. But, in spite of his now elevated position in ADAS, he has retained the down-to-earth approach and puckish sense of humour which made him so popular with farmers and which are both evident in this excellent book.

Tackling Farm Waste brings together all the things farmers ought to know about slurry and farmyard manure and presents them in a readable and readily assimilated form. It combines comprehensive background information with invaluable reference material.

As well as describing methods of disposing of muck and slurry Kevin Grundey suggests how its manurial values can best be exploited. And at a time when the price of energy-based nitrogen fertiliser is rocketing, that alone should help farmers to avoid wasting a valuable resource and bring some order into a usually haphazard operation.

The more intensive the production, the greater the problems of dealing with waste, and the book shows too how ever-increasing numbers of livestock are being kept on ever-decreasing numbers of farms. It also lists the incredible number of anti-pollution laws, although Kevin Grundey concedes that not all are yet policed. 'But times are changing', he says. 'Each year a little less land is farmed by fewer farmers, so the urban voice grows that much stronger'.

Indeed, there can seldom have been a more opportune moment to publish such a work and it should perhaps be required reading for anyone running, or contemplating starting, an intensive livestock unit.

But if, having followed all the advice contained in these pages, a farmer still suffers complaints from irate neighbours, Kevin Grundey has one final recommendation. 'Tidy the place up and then invite them round', he says. 'Some small refreshment—not too lavish—to complete the tour provides an opportunity for people to talk. We all complain most against unknown people, but not nearly so forcefully against those we have met a few times'.

Altogether a very practical and realistic approach, I'd say!

White Rails Farm, DAVID RICHARDSON
Great Melton,
Norwich

AUTHOR'S NOTE

THIS BOOK is for farmers, not scientists: in saying this, the intention is not in fact to give offence to either party, but to recognise that the farmer must run a business which is a dynamic and demanding operation. There is not the time to delve into great detail on every topic which farm waste seems to cover.

Thus this book deals in greater detail with those aspects more directly under the farmer's control such as choice and operation of systems. There is in addition what is hoped will be useful information of a reference nature. Rash indeed would be the author who believed his work would provide 'all the answers' but at least this work ought to start the farmer reader thinking—and therein cultivate the farmer's real 'real estate'.

To other readers, perhaps the wide-ranging nature of farm waste may come as something of a surprise.

ACKNOWLEDGMENTS

THIS BOOK would not have been produced without the persistence of Mr Philip Wood of Farming Press, who has proved to be the most helpful of publishers. My debt is equally real to my wife Beryl who not only typed the work and carried out all the secretarial duties, but also offered considerable encouragement—convalescence from which is almost complete!

My thanks are also due to Mr W. Dermott, Agricultural Development and Advisory Service, Ministry of Agriculture Fisheries and Food, without whose permission I could not have written. Last, but by no means least, to the many farmers and colleagues who, over the years, have provided challenge and stimulation in grappling with a wide range of advisory problems thrown up by that most fascinating of industries, agriculture, must go my unremitting thanks.

Caversham, Reading.
February, 1980

K.G.

Chapter 1
INTRODUCTION—
Basic Problems with Farm Waste

THE MOST fundamental problem is what shall be regarded as farm waste? A glib definition of waste is surplus or unwanted material. Quite possibly at another time or place it may have a value, and the idea of re-using waste is gaining strength in these energy conscious times.

In the farming world, judging from enquiries received, 'farm waste' seems to include old drums and spray containers, straw, dead animals, scrap metal, old hay, surplus spray chemical mixture, old oil, surplus milk and whey, foul smells, afterbirth, crop residues, washing down and other waste waters, in addition to the many aspects of manure production, handling, storage and utilisation. In short, almost anything which becomes a nuisance on the farm.

In attempting to tackle farm waste, which as a topic sprawls over so many subject areas, the farmer must acquire information from many disciplines, become competent with a variety of technology and take on the mantle of engineer, microbiologist, designer, builder and scientist, as well as develop more than a passing awareness of the minefield of relevant legislation. To do all this as a mere sideline to running a farm, is something of a tall order. But the need is real because of the many pressures on agriculture.

There is the pressure of steadily reducing land area for agriculture. At present the annual rate of loss is about 25,000 hectares. As a result more urban dwellers are brought into contact with farming, and this is aided further by the freedom to travel brought by the privately-owned car, so much more widely distributed in the population nowadays. There is, moreover, a great deal of interest in environmental quality, especially of the countryside.

Meanwhile, agriculture faces economic pressures being a 'price taker' not a 'price maker'. This has been reflected in two ways. A rapid reduction in the agricultural labour force such that nearly 70 per cent of farms now have no employed labour. At the same time there has been the need to boost financial turnover by keeping more animals. Figures 1–4 illustrate these trends over the past decade. Hence, modern livestock production systems have evolved whereby large numbers of animals are kept in the minimum space compatible with good performance. With the need to cut labour requirement, little or no bedding is used resulting in the handling of the manure as a liquid or semi-solid slurry which often generates a foul smell.

In contrast, when herd sizes were small, the disposal of manure was not a problem, if one chose to disregard the hard and unpleasant physical labour needed in those far-off shovel and wheelbarrow days. Most animals were bedded on straw and the manure handled as a solid. If land was not available for immediate application of the manure, a storage heap was built. Whilst some loss of plant nutrients from the muck heap was inevitable, the aerobic conditions of storage allowed some composting action with the accidental benefit of controlling obnoxious smells. Indeed, the smell of well rotted farmyard manure came to be regarded as 'healthy' and a genuine 'country smell'.

AVOIDING WATER POLLUTION

Another contributory pressure on agriculture is the need to avoid pollution of water supplies. As the national standard of living

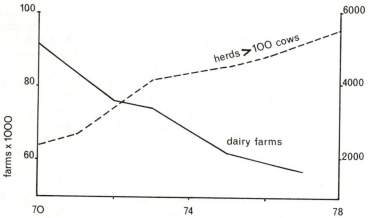

Fig. 1. Dairy cow population 1970–78 (England and Wales).

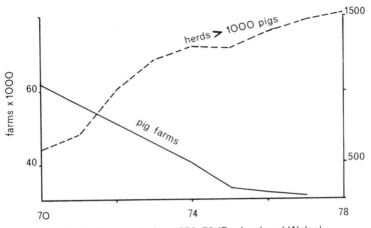

Fig. 2. Pig population 1970–78 (England and Wales).

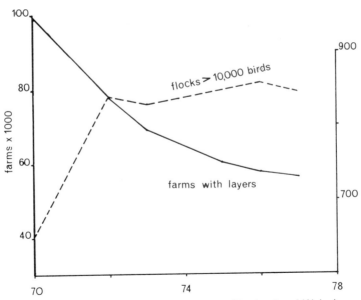

Fig. 3. Laying hen population 1970–78 (England and Wales).

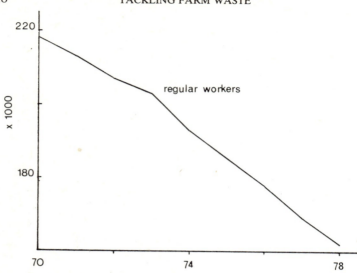

Fig. 4. Farm labour force 1970–78 (England and Wales).

has gone up over the decades, so has the consumption of water for both domestic and industrial purposes. Farming operates on top of the nation's water catchment areas; and as an integral part of food production, farmers need to use fertilisers, spray chemicals and apply manures back to the land—manures which of themselves are pollutants 50 to 200 times stronger than untreated domestic sewage. This need to apply pollutants in the water catchment area, allied to increasing water demand, inevitably focuses attention on the need to avoid pollution of water. As Table 1 shows, however, few farmers are prosecuted for serious water pollution.

Thus livestock farmers are pressured from many sides. Economics and labour shortage dictate more intensive livestock units which produce more manure in a more smelly form; more non-farming folk are interested in the countryside environmental quality, and the need to avoid water pollution grows stronger. In addition, existing legislation to prevent pollution may shortly be strengthened by the pending Control of Pollution Act 1974. These pressures are the root problem in grappling with farm waste.

Many of the potential problems can theoretically be solved by technology but not within the economic framework of farming costs and returns. Herein lies a difficult choice: should the polluter

BASIC PROBLEMS WITH FARM WASTE

TABLE 1. Farm Waste Warnings and Prosecutions 1977

Type of waste	Warnings Total	%	Prosecutions taken	Convictions resulting
Overflowing or bursting of banks of manure stores	283	22	11	8
Silage effluent	195	15	12	7
Run-off from land	144	11	7	5
Dairy and yard washings	318	25	2	2
Pesticides	34	3	1	1
Vegetable washings	22	2	—	—
Miscellaneous	297	23	9	5
Total	1,293		42	28

Source: Water Authorities

always pay to rectify or prevent his own pollution? Most people would see this as fair. However, if the polluter cannot pay, what then? Presumably due to court action by complainants, his business would either have to close immediately or gradually become non-viable as the extra costs of pollution prevention took their toll of profits.

DANGER OF UNDULY STRICT CONTROLS

The closure of the odd livestock unit or two probably would alarm no one other than the farm people involved, but the danger is that precedents may be set which over a period could result in strict controls over livestock production. If these controls became unduly tight or restrictive and pushed up the costs of animal production to uneconomic levels, then farmers would have to obey market forces. At present UK agriculture produces about 60 per cent of our food needs and nationally it would be no particular help to our balance of payments to have to import more meat.

But surely this is an extremist view of the effect of pollution prevention measures upon agriculture? Whatever opinion is held on this, the fact remains that livestock units *are* being closed down either as a result or anticipation of court actions and proposed expansions are curtailed.

As Table 2 shows, nearly 50 per cent of the pig industry resides on 1,400 farms, so we cannot spare too many sites before significantly influencing the industry as a whole. Thus great responsibility

**TABLE 2. Extract from MAFF 1977 Census Data.
Distribution Table for Pigs**

	Herd Size (nos. of pigs)					
	under 200	200–399	400–999	1,000–4,999	over 5,000	Total
No. holdings	22,327	2,994	2,850	1,367	69	29,607
% of total	75	10	10	5	0·2	100
No. pigs, '000	983	857	1,789	2,424	509	6,561
% of total	15	13	27	37	8	100

falls on all parties to preserve a judicious balance of interests: planners to plan with broad vision; the public to pursue only genuine complaints; farmers to always use the best practicable means of controlling potential nuisance or pollution from their activities. No one has the right to pollute the environment to the detriment of others, yet we must all eat and drink, work and enjoy recreation (in which enjoyment of the countryside is very important). Compromising these needs is a difficult problem, which can only be settled by society itself.

In this connection, the recent (Sept 79) seventh Report of the Royal Commission on Environmental Pollution in Agriculture, running to 274 pages, makes interesting reading with its 81 recommendations. It recognises the special difficulties of farmers in financing pollution control measures and raises the question of providing aid. More research and development work on control measures is urged as well as more advisory input from the state extension service (Agriculture Development and Advisory Service of MAFF) to assist the agricultural industry. Greater control measures on intensive livestock farming are proposed and farmers' responsibility to avoid pollution is stated unequivocally.

At present, the Government's reaction to the Report is awaited. Inevitably, implementation of the recommendations would cost money and nationally we are in straitened financial circumstances. It will be an unenviable task to decide priorities and what can be accomplished.

MANURE PRODUCTION

The largest part of farm wastes is the manure produced by animals, and Table 3 shows the national livestock population, manure

TABLE 3. Excreta Production by Livestock in England and Wales
(1978 Agricultural Census Data)

Livestock	No. (million)	Total excreta (million tonnes/yr)	Theoretical value of available nutrients (£m/yr)	Excreta produced by housed livestock (million tonnes/yr)	Theoretical value of available nutrients (£m/yr)
Dairy cows	2·7	41	60	20	30
Other cattle and calves	6·9	43	64	21	32
Pigs	6·5	9·5	18	9·5	18
Poultry	113·0	4·7	21	4·7	21
Sheep and lambs	21·3	21·7	35	—	—
Totals	150.4	122	£199	55	£101

Footnotes: all figures rounded
all cattle assumed housed 180 days

produced and its theoretical value. That there is a considerable financial bonus to be earned by utilising manures effectively to save purchased artificial fertiliser is very clear.

The MAFF annual agricultural census produces data to show, amongst other statistics, the livestock population of the fifty-five counties in England and Wales and even down to parish level (there are about 14,000 parishes). Other tables, known as 'frequency distribution tables', show the size structure of sectors of the industry. These tables show for a given size group of holdings, the number of animals kept and the size of herd. Space precludes quoting such a table in full but an extract of the data showing the breakdown of the national pig herd is illustrated in Table 2.

It is not generally known that much of the census data is available for purchase (a few pence for a table). A price list may be obtained from MAFF Agricultural Censuses and Surveys Branch, Government Buildings, Epsom Road, Guildford, GU1 2LD.

Chapter 2

METRICATION AND GLOSSARY OF TERMS

AT PRESENT the UK is only part way through the conversion from imperial units of measurement to metric terms. UK agriculture has made considerable progress in converting everyday commodities to metric, such as kilograms of feed or litres of milk, but there is still some way to go yet. Whilst fierce argument may rage over the advantages or otherwise of 'going metric', it is now a fact of life and the sooner the pain barrier is broken through the better for each individual.

The worst mistake is to measure distances in imperial and then have to convert to metric. By measuring in metric from the beginning of a problem, all the considerable benefits of the decimal system are available immediately. After a little practice the benefits become very clear. One of the first moves is to re-calculate the acreages of the fields on the farm map into hectares.

It is no function of this book to be a comprehensive source of metric conversions, but for the reader's convenience a list has been drawn up of those quantities which are most likely to be encountered in dealing with farm waste; for most purposes the approximate equivalent quoted will be enough. The metric system is based on the decimal or unit of ten and the general terms to describe the various units are as follows:

DECIMAL MULTIPLES AND SUB MULTIPLES

Prefix	Symbol	Factor by which unit is multiplied	
tera	T	1 000 000 000 000·0	(10^{12})
giga	G	1 000 000 000·0	(10^9)
*mega	M	1 000 000·0	(10^6)
*kilo	k	1 000·0	(10^3)
*hecto	h	100·0	(10^2)
deca	da	10·0	(10^1)

deci	d	0·1	(10^{-1})
centi	c	0·01	(10^{-2})
*milli	m	0·001	(10^{-3})
*micro	μ	0·000 001	(10^{-6})
nano	n	0·000 000 001	(10^{-9})
pico	p	0·000 000 000 001	(10^{-12})
femto	f	0·000 000 000 000 001	(10^{-15})
atto	a	0·000 000 000 000 000 001	(10^{-18})

These are likely to be the main prefixes used in agriculture and horticulture.

Examples are:
- MW = megawatt
- km = kilometre
- hl = hectolitre
- cm = centimetre
- ml = millimetre
- μm = micrometre
- acres × 0·4047 = hectares (ha)
- ft × 0·3048 = metres (m)
- ft/min × 0·0051 = metres per second (m/s)
- ft^2 × 0·0929 = m^2
- ft^3 × 0·0283 = m^3
- gal × 4·546 = litres (l)
- gal × 0·004546 = m^3
- gal/ac × 11·234 = l/ha

- gal/min × 0·0758 = l/s
- horsepower × 0·7457 = kW
- inches × 25.4 = mm
- in^2 × 645·2 = mm^2
- in^3 × 16,390 = mm^3
- lb × 0·4536 = kg
- lb/m^2 × 6·895 = kilopascals (kPa)
- miles × 1·609 = kilometres (km)
- ton × 1·016 = tonne
- tons/ac × 2·511 = tonne/ha
- units/ac × 1·255 = kg/ha
- yd × 0·9144 = m
- yd^2 × 0·8361 = m^2
- yd^3 × 0·7646 = m^3

1,000 litres = 1m^3 ⎫ for slurry taken as equal
1,000 kg = 1 tonne ⎭ *i.e.* 1 kg = 1 l; 1 m^3 = 1 tonne.

GLOSSARY OF TERMS

Like most subjects, farm waste has its own jargon and shorthand. Such abbreviations are often criticised as being a sort of defence mechanism used by the knowledgeable types to make understanding the subject more difficult for those who want to learn. No doubt there is a grain of truth in this on occasions, but abbreviations do make reading and comprehension easier, especially where characteristics or measurements are referred to repeatedly. The eye would surely fatigue if abbreviations were not used. To read that 'the biochemical oxygen demand (5 days at 20°C test method) of pig slurry is 25,000 milligrams per litre,' is easier taken in if presented as 'pig slurry has a BOD$_5$ of 25,000 mg/l.' Thus this section on glossary of terms is deliberately placed early on in the book not to deter the reader but simply to draw attention to its presence.

It is a matter of personal choice whether to read through the abbreviations and their accompanying explanatory notes or to move on, returning to check out abbreviations as they arise in the following chapters. The notes are necessarily cryptic to avoid overburdening the reader with 'scientific' information which can be more fully and accurately obtained elsewhere. To this end, a few references for further reading are added at the end of the glossary for those seeking greater detail.

AEROBIC—In the presence of free oxygen. Usually used in reference to conditions required for some types of bacteria to thrive. Oxygen held within the make-up of a compound is not available for aerobic bacteria unless the compound is broken down to release the oxygen.

ANAEROBIC—Conditions from which oxygen is absent. Strict anaerobes will not exist in the presence of oxygen although some bacteria can adapt to aerobic or anaerobic conditions. (See facultative bacteria).

BOD_5—The amount of oxygen taken up by an organic waste or pollutant during a standardised test. The sample is held under specified conditions *e.g.* 20°C for five days and its oxygen demand measured using laboratory procedures. The units quoted are milligrams of oxygen per litre of sample, mg/l (though much less than a litre sample is used). The BOD_5 figure is a measure of the more readily degradeable organic matter in a material and so reflects its polluting effect on entering a river. The higher the number, the greater the polluting power. Note that this BOD_5 figure is quite different from the COD figure.

CADMIUM (Cd)—A heavy metal present in some sewage sludges. It accumulates in the soil and is taken up by plants to the detriment of the consuming animal or human. Although low levels of Cd are suggested as tolerable, investigation continues on Cd levels in soils and crops. Many farmers avoid using sewage sludges containing Cd.

CARBON DIOXIDE (CO_2)—A poisonous gas, heavier than air so collects in below-ground chambers or tanks. Released, with other gases, by fermenting manures and vigorously so, if silage effluent is mixed with slurry and stirred.

CHEMICAL OXYGEN DEMAND (COD)—Determined in the laboratory by boiling for two hours a sample of waste or polluting material mixed with sulphuric acid and potassium dichromate, to determine the total amount of oxygen needed to oxidise *all* the organic content. The test is much quicker and produces a higher reading than the BOD_5 test. Because of its speed and repeatability, water authorities tend to prefer the use of COD.

As a measure of polluting power, COD probably overstates the case where mild pollutants are discharged to running water, since complete oxidation by the biology in the stream or river may take months. In the UK where rivers are short, the pollutant will have reached the sea long before complete oxidation can occur.

COMPOSTING—The maintenance of solid organic waste material in an aerobic condition enabling digestion and breakdown of the waste to convert it into a usually fibrous dark brown not unpleasant smelling damp material. Well known to gardeners, the process has been investigated at laboratory and pilot scale for the treatment of pig slurry, and patents have been secured by the National Research Development Corporation. This is the body responsible for patenting and commercialising new processes discovered in government financed institutions such as research institutes, universities, etc.

COPA—Control of Pollution Act 1974. Has six parts of which Part 2 (Prevention of Water Pollution) will affect farming more than the other five. Originally postponed on its inception because of the costs involved, several parts of the Act are now in force. The whole of Part 2 of the Act was to have been implemented at the end of 1979, but this is now quite uncertain.

COPPER (Cu)—Added to pigfeed to stimulate growth and so excreted in pig faeces. Too much pig muck applied to grazing can put the grazing animal at risk from copper poisoning. Sheep are most likely to be affected. Too little copper in the diet of young cattle can result in their failure to thrive.

DENITRIFICATION—The reduction or breakdown of nitrates in a waste accompanied by the production of nitrogen (N) as a gas. Traditionally, loss of N is regarded as undesirable as this is the most valuable fraction of the fertiliser value of a waste.

DIGESTION—Used to describe a chemical or biological process in which compounds or materials are broken down to form other simpler compounds.

DM—dry matter. That fraction of a substance left after all water has been removed, usually by heating to 105°C for twelve hours in a laboratory oven. Same commodity also referred to as 'total solids'.

EUTROPHICATION—The enriching of natural waters with plant nutrients, usually N or sometimes P, thus enabling the growth of aquatic plants to be enhanced. The practical problem is that when the increased plant life dies due to seasonal effects or overcrowding, the decaying organic matter exerts an oxygen demand on the water and may even exhaust the natural oxygen supply, causing the death of fish and other aquatic species. The living or dead plant material may also block drains, pumping inlets, etc. Eutrophication is of great concern to reservoir managers in their quest to maintain high standards of purity in drinking water.

FACULTATIVE BACTERIA—Those which can exist in the presence or absence of oxygen.

FARMYARD MANURE (FYM)—A mixture theoretically of animal faeces, urine and bedding, usually straw. In practice, many other commodities may be present (see also slurry). FYM is a material which can be handled as a solid and is stackable having a DM ranging from 30 to 50 per cent. One cubic metre (1 m^3) is reckoned to weigh 0·9 tonne, although weight will vary with the amount of straw present and whether the manure is settled and compacted or freshly disturbed.

MAGNESIUM (Mg)—A trace element necessary for healthy plant growth. Grazed grass too low in Mg may give rise to the disease hypomagnesaemia in cattle, which can be fatal. In practice this may arise where excessive amounts of potassium are applied to the crop. The herbage takes up extra potassium and the necessary equivalent extra magnesium may not be available.

NIAE—National Institute for Research in Agricultural Engineering, Wrest Park, Silsoe, Bedford. An Agricultural Research Council financed centre of research.

NITRATE (NO_3)—A chemical compound or radical which exists by bonding onto other groups. A simple example would be ammonium nitrate, a common artificial fertiliser, but in farm wastes the NO_3 may be bound up with organic compounds. Greatest interest and concern lies in the problem of excess nitrogen application to land resulting in NO_3 being washed from the soil into the drains and eventually reaching drinking water supplies. The World Health Organisation has laid down certain standards suggesting limits to the amount of NO_3 in drinking water. (A maximum of 22·4 mg/l of nitrate nitrogen (NO_3) and nitrates (NO_2) are thought to be implicated in the occurrence of cancer.)

NITRIFICATION—The breakdown of organic material to form NO_3 and NO_2 by nitrifying bacteria. This resulting NO_3/NO_2 may be further broken down (denitrified) to release N as a gas. Simultaneous nitrification and denitrification can be achieved by manipulating conditions within a waste treatment process.

NITRITE (NO_2)—A compound usually unstable in that it is formed often when NO_3 is being built up or broken down. Of rather academic interest in practical farming conditions.

NITROGEN (N)—The most important fertiliser element. It is present in most farm wastes and is the most valuable element when looking at the financial value of manures as fertiliser for a crop. Usually combined with other elements as NO_3 or ammonium compounds, N can be lost as a gas from manures due to digestion process during storage or when manure is left lying on the field surface.

pH—A logarithmic scale 0 to 14 measuring the acidity or alkalinity of a liquid. pH 7 is neutral, below 7 acid and above 7 alkaline. Farm wastes tend to be in the range just acid to just alkaline with the notable exception of silage effluent which is very acid with a pH around 4.

PHOSPHORUS (P)—One of the three main fertiliser elements. Regarded as being held tightly by soil particles, some soils will lose small amounts into drainage water. With short rivers in the UK, P is not usually regarded as a troublesome water pollutant. However, where rivers drain to land locked water, phosphate levels steadily rise as small amounts of leached phosphate accummulate and can be of great concern over eutrophication of the water. A good example of this is Lough Neagh in Ireland and some reservoirs in the UK.

'P' content of fertiliser may be declared as 'P' or 'P_2O_5', the relationship being $P \times 2.29 = P_2O_5$.

POTASSIUM (K)—One of the three most important elements in plant fertiliser. Potassium content may be declared as 'K' or 'K_2O'. The relation is $K \times 1.2 = K_2O$.

ROYAL COMMISSION STANDARD—A Royal Commission on sewage disposal began in 1898 and continued spasmodically until 1915. One of its recommendations (8th Report, 1912) was that an effluent discharged to a water course would not be regarded as polluting if the standard of the effluent was not worse than 20 mg/l BOD_5 and 30 mg/l SS, and that it was diluted at least eight times its own volume with clean water at the discharge point, *i.e.*, the BOD_5 of the stream should not increase by more than 2·5 mg/l nor the TSS by more than 3·75 mg/l. This standard is sometimes referred to as the '20/30 standard', or 'the Royal Commission Standard.' In broad terms this permissible discharge is about one-tenth or less of the strength of untreated domestic sewage, which is usually in the range of 200–300 BOD_5.

SLURRY—A mixture containing animal faeces and urine plus some of the following: bedding, wasted food, spilled drinking water, cleaning/washing water and rain water. There may also be foreign objects present including wood, stones (brought into the building in cows' feet), bricks, nuts and bolts, spanners and larger metal pieces, string, wire, plastic materials, paper, cardboard, rag and articles of clothing—indeed, any item in use frequently or occasionally about the farm may 'get lost' into the slurry.

To inexperienced readers, the foregoing may appear to be exaggeration yet it is quite likely that experienced livestock farmers could add to the list. In planning slurry handling systems, the possible presence and the effects of such objects is surprisingly often overlooked. The prudent designer will make provision for foreign objects which he will regard as inevitable, remembering that most slurry collection channels and pits are below ground and that gravity seems to exert a strong pull on any unsecured object close to such pits.

SODIUM (Na)—A comprehensive analysis of a waste may declare some sodium content in mg/l. This along with other elements like calcium or potassium forms part of the 'dissolved salts' picture of a waste, mainly of scientific interest. Very high salt levels may create problems in trying to treat a waste.

SOLIDS RETENTION TIME (SRT)—The length of time that a waste remains within a treatment system. In an activated sludge treatment system, the minimum SRT is the doubling time of the beneficial micro-organisms in the system.

TOTAL SUSPENDED SOLIDS (TSS)—Material in suspension in an effluent. The suspended solids (SS) are those which can be recovered by filtration or centrifuging.

TS—total solids, same as dry matter.

VOLATILE FATTY ACIDS (VFA)—Those organic acids in wastes which can be removed by steam distillation. It is these acids which are important in the production of methane gas or biogas when manures, etc, are digested under anaerobic conditions. VFAs are formed in the first stage and are the substrate upon which methanogenic bacteria feed to produce methane.

VOLATILE SOLIDS (VS)—That portion of the total solids of a sample which can be driven off by heating to 600°C for one hour.
ZINC (Zn)—A metal found in small amounts in some farm wastes. Cattle excreta is low but pig and poultry may have appreciable amounts, *e.g.*, 0·5 mg/l. Like all metals it accumulates in the ground, which is to be avoided lest it builds up to levels where it may become toxic to plants. Most crops seem to tolerate up to 250 mg/l in the top 150 mm of the soil.

[REFERENCES]

Studies on Farm Livestock Wastes 1976, *Agricultural Research Council*.
WISDOM, A. S. The Law of Rivers and Water Courses 1980.

Chapter 3

LEGISLATION AND FARM WASTE

OTHER THAN some law students, few ordinary readers will turn avidly to a chapter on legislation. Indeed to place such a chapter so early in the book may be regarded as a positive disincentive to the reader. Many experienced farmers can claim to have survived over the years with little knowledge of the various laws, so why are they so important now?

It seems trite to say that 'times are changing' but in respect of pollution the old adage is so very true. A number of related aspects have come to prominence in the last few years, so that public interest in the environment is increasing noticeably. The framework within which industrial (including farming) practices are considered acceptable from an environmental viewpoint gets tighter as time goes by. Each year there is a little less land being farmed and fewer farmers, so the urban voice grows that much stronger. The two unseasonally dry years of 1976 and 1977 gave the population a shock: for the first time, over large areas of the UK, water supply and quality were in some doubt. This directed attention to water quantity and quality in turn leading on to pollution avoidance.

Very rarely is pollution a deliberate act, most incidents arising from accidental or careless circumstances or even from ignorance of regulations. However, none of these reasons is a defence against a charge of pollution, and in certain situations, serious pollution can result not only in sizeable fines in court, but also perhaps the requirement to cease polluting forthwith. On occasion this has caused a farming enterprise to be closed down for a period—a very costly exercise indeed. The time has come when some knowledge and awareness of pollution and other legislation relevant to farming is a very necessary element of good business management. To be caught unaware could well savage the farm's profit figures. Over and beyond mere financial considerations, there remains the moral aspect that none of us should pollute our environment; we all must eat and drink and doubtless expect uncontaminated foodstuffs as well as a pleasant environment.

For all these reasons, it was thought justified to include a chapter on legislation, and by placing it early in the book, to draw the attention of readers more positively to its presence. The various acts are listed randomly since their importance will vary depending upon the individual farm's circumstances. It is notoriously difficult, even unwise, to attempt to precis legislation since it is almost inevitable that inaccuracy creeps in. However, to merely list the many acts potentially involved in tackling farm waste in practice would be less than helpful, and so an attempted thumbnail of the Acts follows. By no means should this section be regarded as an authoritative explanation; that is the province of the lawyer and ultimately the courts.

1. Rivers (Prevention of Pollution) Acts 1951 and 1961

2. Clean Rivers (Estuaries and Tidal Waters) Act 1960

Taken together these Acts make it an offence to discharge any noxious pollutant to a watercourse which includes any river, stream or ditch, even if the latter runs only part of the year and is dry for the rest.

Farm drainage is classed as a trade effluent and before this can be discharged into a watercourse, the consent of the water authority is needed. Farm drainage, in the context, means any drainage from yards or from field drainage pipes and surface run-off. The conditions of consent may specify the effluent characteristics, the volume and timing of discharge and its temperature. If a consent is applied for and rejected, appeal can be made to the Secretary of State.

Thus it becomes clear that consulting the water authority about any planned changes on the farm which may effect the local water situation is eminently sensible since they are the final arbiters. These acts will be replaced and expanded by the Control of Pollution Act 1974 whenever it may be implemented.

3. Water Resources Act 1963

Section 72 makes it an offence to discharge liquid waste underground by any well, bore-hole or pipe without the written consent of the water authority. An example would be the presence on the

farm of an old pitshaft. This could not then be used to dispose of any farm waste of which both slurry and surplus milk have in the past been dumped down shafts. This act will be replaced by COPA 1974.

4. Salmon and Freshwater Fisheries Act 1975

No liquid or solid matter may be put into water populated by fish which might injure the fish, their spawn or the food of fish. The use of electrical devices, explosives or poisons for fishing is also outlawed. Prosecution is reserved mainly to the water authority, although the secretary of an angling club may be reckoned to have 'a material interest' in the waters affected and so be able to prosecute.

Whilst this may appear to cover much of the provisions of paragraphs 1 and 2 above, in practice it means that water authorities may seek conviction of a polluter under this act as well as others and so strengthen the case-lore against a polluter.

5. Public Health (Drainage of Trade Premises) Act 1937

6. Public Health Act 1961

Part 5 of the 1961 Act brings farm drainage within the meaning of a 'trade effluent'. This gives farmers, along with other industries, under the 1937 Act, the right to discharge effluents into the public sewer system. This right is subject to the consent of the water authority, who are empowered to stipulate quality and quantity restrictions and to charge for treating the effluent.

With the generally overloaded state of many sewage works, the full cost of treating farm effluent will usually have to be met by the discharging farmer. In view of the strength of farm wastes, these costs will be high and usually will make sewer disposal economically unattractive.

The National Farmers' Union and the National Water Council on behalf of water authorities have drawn up a model agreement and formula to fix costs.

At present it is not known how many farms are connected to public sewers but estimates have put this at less than 4,000 farm premises.

7. Deposit of Poisonous Waste Act 1972

It is an offence to deposit on land any poisonous or polluting waste in such a way as to cause danger to persons, animals or to a water supply. Putting the waste in containers and depositing the containers on land would not escape this law.

8. Public Health Act 1936—also amended by the Local Government Act of 1972

Local authorities can make bye-laws to prohibit any keeping of animals which is prejudicial to health. Other bye-laws may be enacted to control the effects of 'noxious trades' which are defined in the Act or may indeed be designated by a local authority. Such trades are generally those processing or using animal by-products. Examples are fat renderers, glue works, etc. Agriculture escapes this description.

Section 91 of this Act requires local authorities to make periodic inspections of their district to detect any nuisances such as premises or animals kept in such a way as to provoke health risks or nuisance. Waste accumulations, dusts or effluvia, any pond, ditch, gutter or watercourse which poses a health risk or nuisance is also covered. Having identified an occurrence, the authority must serve an abatement notice on the polluter which if not complied with would entitle the authority to institute court proceedings. If the court finds for the prosecution, then a fine may result and a nuisance order will be served to prevent the nuisance. A polluter continuing to pollute after a nuisance order may be fined again and the court may impose a daily fine too. An alternative is that the authority may correct the nuisance and recover the costs as well as legal expenses from the convicted polluter.

This looks strong medicine for a polluter, but worse follows. If a local authority feels that the above summary proceedings will not provide an adequate remedy, they may, under Section 100, proceed in the High Court to seek abatement or prohibition of the nuisance. Should the polluter then continue, he could be committed to prison because of contempt of court.

One of the most potentially difficult aspects of this Public Health Act 1936 is that *individuals* can also make a complaint to magistrates. Usually, where a complaint (for example, about smells) is substantiated, the local authority will prosecute but only after genuine and lengthy attempts to achieve an abatement between the polluter and complainants. However, in today's financially stringent times an authority may sometimes feel that the severity of the complaint would not warrant the time and effort of going to court. There is nothing to prevent the complainants taking up legal arms and seeking a prosecution, always provided they have the financial resources.

Where a nuisance order is sought in respect of accumulations or deposits, dust or effluvia from a trade, business, manufacture or process which is prejudicial to health or a nuisance, the polluter may use as a valid defence proof that the 'best practicable means' have been taken by him to control the nuisance. There is, needless to say, a number of shades of opinion as to what best constitutes 'best practicable means'. This defence is not available in High Court actions.

9. Public Health (Recurring Nuisances) Act 1969

Where a local authority believes that a nuisance is likely to recur on the same premises, it may serve a prohibition notice to prevent the polluter from repeating the nuisance. They may also specify work necessary to prevent recurrence. A prohibition notice may be served whether or not an abatement notice has been served and regardless of whether or not the nuisance exists at the time of serving the prohibition notice. The latter can be enforced via the magistrates' court.

10. Local Government Act 1972

This empowers local authorities to make bye-laws for the prevention and suppression of nuisances in their areas. Such bye-laws are subject to confirmation by the Secretary of State.

11. Town and Country Planning Act 1971—The Town and Country Planning General Development Order 1977

The 1971 Act stipulates that all development (building works, etc) needs planning permission but the *use* of land or buildings associated with the land for the purposes of agriculture is not regarded as development.

The latter order made under the Act, exempts some types of development, including agricultural ones, from the necessity of obtaining planning permission. Providing building works and erection of buildings is 'requisite for the use of the land', then no permission need be sought as long as the building is less than 12 m in height, less than 465 m^2 in area and more than 25 m from a classified metalled road. Buildings closer to the road, agricultural dwellings and any building on a holding less than 0·4 ha need planning permission.

The local authority can bring a proposed but apparently exempted development under control by reference (known as an Article 4 direction) to the Secretary of State. The authority must have a sound and sufficient reason for so doing and the Secretary of State will consult MAFF on the matter. An example might be a tower silo in an area of outstanding natural beauty.

It is under the planning legislation that farmers intending to develop the larger intensive livestock units, either by extension of existing buildings or by all new erection, will often come up against local authority requirements concerning waste handling systems and arrangements. The consultative process can be very long and involved, sometimes exasperating, but there is little to be gained by expostulation. The only worthwhile precaution is a carefully researched proposal, taking competent advice and incorporating wide discussion with likely interested parties, official and otherwise, *from the earliest possible moment*. On occasions, a year will be inadequate to plan, build and occupy a new unit.

12. Countryside Act 1968

This imposes on government ministers and hence their departments, plus other public bodies, the duty of bearing in mind the need to conserve the amenity of the countryside. This means that any application for grant aid, *e.g.*, FHDS, FCGS, etc, in its examination will be considered from this angle in addition to other financial or technical requirements.

13. Building Regulations 1976

These lay down standards concerning structural and design aspects of buildings. The regulations are updated more frequently than many other acts and may be further reinforced by local authority requirements too. Since building regulation compliance is necessary for building work, especially drains, sewage and sanitary matters, it is necessary for a local authority inspector to visit the site several times. It is by the application for inspection, etc that a local authority gains knowledge of proposed building work and so it is not possible to avoid the need to apply for planning permission.

There have been odd instances, where the more piratically minded have short circuited the whole planning and regulation system by simply not applying for any permission whatsoever and erecting the new building. Nowadays, this sort of anti-social behaviour is also uneconomic! Following discovery, to have to dismantle the new building to reveal foundations, etc for subsequent satisfactory inspection can be disruptive and expensive in fees for the inevitable legal battle which cannot be won by the transgressor.

14. Control of Pollution Act 1974

Passed on 31st July 1974, this Act is wide-ranging in its endeavour to improve control of pollution and nuisance. It will also supersede various existing anti-pollution legislation. Because of costs involved, much of the Act was not implemented at once and many sections yet remain in abeyance. There are six parts:

Part 1—Waste on land.
Part 2—Pollution of water.
Part 3—Noise.
Part 4—Pollution of the atmosphere.
Part 5—Supplementary provisions.
Part 6—Miscellaneous.

Agriculture is principally affected by Parts 1 and 2, although Parts 3 and 4 may be of some concern. Parts 5 and 6 deal with procedures, appeals etc.

Part 1 (Implemented partly in 1976 and partly in 1978)

This establishes the concept of Waste Disposal Authorities (WDA) as the responsibility of county councils. They must be responsible for the collection and safe disposal of 'controlled' waste. This is defined as waste from household, industrial and commercial premises. Agricultural waste is excluded from these provisions. The WDA must find out the nature and volume of waste arising in their area, including farm waste, so that they can adequately plan for its disposal.

The main control is achieved by prohibiting the dumping of controlled waste on land other than at a licensed site. Farm manures are not involved here, although there are some opinions which question the status of a manure lagoon.

Part 2. *(Mainly yet to be implemented)*

It is this section of COPA which is likely to have the most influence on agriculture. It specifies and seeks to protect from pollution 'relevant' waters, *i.e.*, coastal waters and estuaries, streams and any underground water which may be specified by the water authority as being underground water for the purposes of the Act.

It was realised that agriculture is a special case here, in that it is established practice to apply polluting wastes (manures) to land as part of the production of food. Yet the land used—farming's factory floor—is also the catchment area for the nation's drinking water.

This dilemma is to be resolved by the concept of a Code of Good Agriculture Practice (GAP). In principle, if a farmer is charged with having caused water pollution, then it will be a sufficient defence for him in the lower courts if he can show that the pollution occurred whilst he was operating in accordance with GAP. That a particular practice may not be specified in the Code does not prevent its being classified as good agricultural practice and so available as a defence. This paragraph is to allow for changes in agricultural practices over the years to be accommodated.

To apply manures or fertilisers to land in a way which is not within the Code of GAP is not of itself to be regarded as an offence. The significance is that if, as a result of that application technique, the farmer were prosecuted for pollution, then he could not invoke the protection of GAP. Clearly, he would be in a poor position tactically in the court. The merit of adhering to the code of practice becomes obvious, taking the view that constraints on pollution prevention will get tighter aided by greater public interest in environmental quality.

Drawing up the code of GAP is the responsibility of the Minister of Agriculture in consultation with interested bodies. No such codes have yet been published, since the implementation date of Part 2 set for 31st December 1979 has been put off indefinitely as part of the Government's economy measures—preventing pollution costs money and resources.

The concept of the GAP as a protection for farmers is sometimes criticised by the un-informed as a bolthole so that farmers can escape the provisions of COPA. Whilst no codes have been published, it is a reasonable expectation that they will reflect what is genuinely accepted as good practice anyway in terms of fertiliser and manure levels for crops. The protection afforded by the code can be withdrawn by an appeal to the Secretary of State by a water authority where it believes that pollution is occurring or is likely to occur as the result of a farmer's operations, even though they fall in with the Code. This may appear to make the Code's protection not worth very much to the farmer, since it may be negated as described.

Such appeals will not be quickly granted because an appeal to the Secretary of State is necessary for each farmer concerned. As part of the process, copies of the water authority application must also be served on the Minister of Agriculture and the occupier of the premises to which the notice applies. The occupier then has 28 days to put his case to the Secretary of State for the Environment, who will then decide the merits of the case. To do this adequately, the advice of the Minister of Agriculture must be sought.

Where in a particular area the Secretary of State believes that the relevant waters may be at risk from pollution, *e.g.*, an area around a borehole, then he may 'designate' the area and prescribe whatever activities need to be controlled specially. There are appeals procedures and a public enquiry may be involved before the control of such 'activities' may be exercised by the water authority. Even then the water authority must not withhold consent to prescribe activities unreasonably, and if necessary, an aggrieved farmer could again appeal to the Secretary of State, who would decide the issue in consultation with the Minister of Agriculture if a Code of GAP is involved.

Other paragraphs of COPA Part 2 prohibit discharging without water authority consent, any effluent into relevant waters or onto or into any land, lake, lock or pond. The discharge of trade effluent is brought more fully under control from July 1976 than provided for under the Public Health Acts of 1937 and 1961. In certain circumstances, water authorities can take steps to prevent pollution injurious to stream life and recover the costs incurred from the person responsible for the pollution.

Part 3 *(Fully implemented in 1976)*

This seeks to control nuisances from noise or vibrations. A local authority can not only require the offender to control the nuisance, but may also serve such a notice *before* a nuisance has occurred. More interesting, is that a complaint to a magistrates' court may be made by only one occupier of premises suffering noise, whereas three or more people were needed under the old Noise Abatement Act 1960 (which is repealed by this Part 3 of COPA).

It would be possible for a single complainant to take action about, say, the noise from pigs at feeding time or of crop-drying fans. Neither of these would be easy to control.

Part 4 (Implemented 1976)

This is really directed at preventing atmospheric pollution by dust and smoke, etc. Not all the required regulations have yet been promulgated and, as written, the Act may cover the problem of straw burning. However, several local authorities already have local bye-laws to regulate burning. The problem of odour is not dealt with by this part of COPA. Complaints about smell are usually taken under the Public Health Act 1936.

Parts 5 and 6 (Mainly implemented 1976)

These enable local authorities to obtain information necessary to discharge their responsibilities under COPA and to enter and inspect premises. In essence the enabling sections of COPA, they also increase substantially the penalties for pollution offences.

15. Health and Safety at Work Act 1974

There are a number of different regulations, *e.g.*, The Agriculture (Safeguarding of Workplaces) Regulations, by which Parliament has attempted to ensure minimum standards of safety at work are maintained in agriculture. Most farmers are well aware of these requirements, and a useful explanatory leaflet, AS1 *A short guide to the 1974 Act*, is available from local Agricultural Inspectors of the Health and Safety Executive giving guidance in some detail. This can always be further supplemented by on-the-farm advice from H and SE staff.

In simple terms the 1974 Act places obligations on:
  (i) employers to ensure employees are not exposed to risks to their health or safety;
  (ii) employers and the self-employed to conduct their businesses

without risk to the health or safety of themselves, other people and especially children;

(iii) employees to take care of their own health and safety as well as other people who may be affected by work activities.

A moment's consideration of these obligations makes it clear how comprehensive they are. Thus it would appear that manure storage installations must be rendered safe not only to workers and the farmer himself, but also to visitors—even to trespassers! Very often, the safety regulations are still looked upon as bureaucratic interference on the farm rather forgetting the objective behind the regulations namely, the prevention of injury, mutilation or even death to real people not statistics. Unfortunately, but understandably, it is only when one has seen at close quarters the grief, pain and recriminations following a serious accident to someone that one becomes imbued with the resolve never to let a repetition occur. One of the reasons why the safety regulations are so important is that they attempt to achieve the same end but *without* incurring the cost of an accident first to drive the message home.

The admonition to 'think first: think *safe*' has much to commend it to us all.

Chapter 4
MANURES AND SLURRIES

VOLUMES OF ANIMAL WASTE PRODUCED

IN DEALING with any practical farm waste problem, the most obvious need is to be able to quantify the amount of waste being produced by the livestock on the farm. It is perhaps surprising to relate that such basic information has only relatively recently become available for many species, and that for some groups of livestock, the figures quoted are results from comparatively few observations or recordings. Perhaps studying in detail the production and quality of faeces is not the most appealing of subjects!

As already mentioned, there is the difficulty of getting figures which are the result of many observations and in which confidence can be placed. This is not to criticise research workers generally; it simply acknowledges that a particular piece of experimental work may require records of faeces and/or urine production but not necessarily of a complete chemical and micro-biological analysis. Thus, to be able to quantify all the various characteristics of the manure from an animal, information from a number of experiments may have to be added together, *e.g.* weight and volume from one set of observations; plant nutrient analysis from another study and micro-biological figures for BOD from a third set of results.

Furthermore, the quantity and quality of manure produced will vary. The size of animal is important as larger ones eat more. Whether or not it is a production animal has a bearing too—a cow in milk is fed more than when dried off and, moreover, different quality rations are needed. Drier, more fibrous rations result in smaller amounts of higher dry-matter faeces—compare cattle fed on hay and concentrates with the sometimes spectacular results when they are turned out to spring grass or get into the kale crop accidentally. The resultant freer-flowing, more copious manure can transform the measured orderly routine of the parlour operation into a potentially dreadful form of Russian roulette!

In the pig world, using swill for feeding may well quadruple the

TABLE 4. Livestock Excreta Production

Type of livestock	Approx bodyweight (kg)	Excreta: litres/day Range	typical	Excreta-approx % dry matter
CATTLE				
Calf—up to 2 months fed conc. liquid feed	73	4·0–6·1	5·0	12–14
Calf—up to 6 months	140	6·3–7·8	7·5	12–14
Heifer—up to 12 months	270		15·0	12–14
Heifer—up to 18 months	380		20·0	12–14
Beef store—up to 12 months	400	10–34	27·0	12–14
Bull	900			
Dairy cow	500	32–54	41·0	13
HORSE	680		30·0	9*
MINK			0·2*	13
PIGS				
Piglet—up to 3 weeks	5		1·0	10
Weaner	12	1·5–2·5	2·0	10
Fattener fed dry meal	50	2·0–5·5	4·0	10
Fattener fed Water:meal				
Fattener fed 2½:1	50	2·0–5·0	4·0	10
Fattener fed 4:1	50	4·0–9·0	7·0	6
Fattener fed swill	50	very variable	15·0	3–5
Fattener fed whey	50	14·0–17·0	14·0	2
Boar			5·0	10
Dry sow			4·5	10
Sow & litter to 3 weeks			15·0	10
POULTRY				
Broiler (incl shavings litter)	2		0·04	60
Duck	2		0·03*	12
Goose	3·5		0·55*	25
Laying hen	2	0·10–0·14	0·114	25
Rabbit	2·5		0·39	—
Turkey	7		0·17*	23

Footnotes—Excreta comprises faeces + urine only.
   —*denotes tentative values.
   —To excreta production must be added estimated amounts of wasted bedding, feed and washing down waters to determine daily waste production.

amount of waste produced compared with pigs fed dry meal and water.

From all of this, it can be agreed that the amount of manure produced by an animal will vary significantly as a result of a number of other factors. If this seems a statement of the obvious, the point will have got home fully about variations. It must be reported, however, that many enquirers about manure quantities seem to take a very poor view when informed that this particular quantity is 'approximately' so much and quite overlook—don't seem to want to know about—the variations. In this modern day and age there is perhaps a subconscious inclination to want exact information, and to admit to an answer which may vary almost smacks of inefficiency, but animals *do* vary in their manure output.

In Table 4, for a number of animals a range of manure volumes is quoted and also a single 'typical' figure. It is this which should normally be used for planning purposes when deciding how much manure storage capacity should be installed. The typical figure is based on experience and has stood the test of time and practical application. In formulating an accurate forecast of manure storage requirements under practical conditions, much depends upon the discretion with which Table 4 is applied.

MATERIALS ADDED TO MANURE

As has already been mentioned, under practical farming conditions a number of additional materials become mixed with the animal excreta.

Bedding materials

The most obvious addition is from wasted bedding and this has a stiffening effect on the slurry by raising the dry-matter content. As more bedding is added, so the slurry consistency changes from very free running to a semi-solid until it becomes a stackable solid, or, in other words FYM. The absorbency of bedding varies and some examples are given in Table 5.

There are in addition a number of other commodities used for bedding livestock and dairy farmers especially have tried a number of variations. Whilst straw tends to be the first and more natural choice of many farmers, lack or high cost of straw in some areas of the West and South West, has stimulated the use of alternatives in cubicles including:

- *Compacted old FYM*. This seems to provide a warm acceptable

TABLE 5. Absorbency of Some Bedding Materials

Bedding material	Water absorbed litres/100 kg bedding
Newspaper, shredded	330
Peat moss	1,000
Sand	25
Sawdust, pine	250
Straw, wheat	220
Straw, barley	200
Wood chips, pine	300
Wood shavings, hardwood	150
Wood shavings, softwood	200

bed provided it is kept dry. A free-draining base to the bed is essential such as rammed chalk, crushed limestone or railway ballast topped with finer material. The reservation about FYM is that it can easily become a warm damp breeding ground for bacteria, and mastitis is never far away.

- *Newsprint.* Shredded newspaper, but not glossy magazines, is quite suitable for cows and can be purchased in baled packs. Unfortunately, the price is not cheap and can vary from time to time.
- *Sand.* Of recent times this became the fashion for cubicle bedding. Great advantages were discovered for sand such as cheapness (only a tonne a winter for a cow), ease of handling, but above all cleaner cows and a bed micro-climate very discouraging to disease. There have been reports of abrasions on cows where hard sand was used and there have been tales of woe amongst liquid slurry handling systems which didn't enjoy a sand diet one bit—blocked pipes, worn pumps and tanks developing sand bottoms a metre deep which could be extracted only with great difficulty.
- *Sawdust and shavings.* Although micro-biologists can discover a great deal of bacterial life in sawdust beds, cows and dairy farmers seem to be in favour. Supply continuity and price can fluctuate but where a good working arrangement to handle in bulk is established, costs can be acceptable. About 4 kg a cow per week are necessary. Given good slurry store management, sawdust doesn't seem to be a problem in the manure.

Shavings aren't universally popular. They can blow about and can form wads to block drain covers, etc. Wood is very slow to

TABLE 6. Straw Bale Characteristics

Bale type	Weight (kg)	Dimensions (m)	Density (kg/m^3)
Pick-up	18	0·36 × 0·46 × 0·99	110
Big round	345	1·7 diam × 1·55	98
Big square	325	1·5 × 1·5 × 2·33	62

break down when applied to land, and so shavings can lie about fields a long time if the muck and shavings is not cultivated into the soil.

Because of costs, there is a need to restrict the amount of bedding used to the minimum compatible with clean and comfortable animals. An additional and very important reason for less bedding is that long, staple material can be a real headache in slurry systems because of the blockages which the lumps of matted fibres can cause. Just how much bedding to use is the subject of great arguments, which most good livestock operators will ignore, being content to keep their stock in their own way.

Cows kept in cubicles might use about a third of a small pick-up bale per week each, around 150–180 kg over a six-month winter. In contrast, cows loose housed in roofed-over strawyards would need at least one and a half bales per week or 700–900 kg each per winter. In the past bedding straw-yards was often seen as a laborious task. However, the advent of large round or square bales containing 350 kg or more of straw (Table 6) and the associated mechanical handling equipment has radically changed this task to an acceptable weekly chore for one man who may take half an hour to put three or four bales into a yard of fifty cattle and tease out the fresh bedding. To do the same job by hand with pick-up bales would take three times as long.

POULTRY LITTER

The battery cage system uses no litter and has supplanted the old system of keeping laying birds on deep litter. The broiler system of producing chicken meat for human consumption uses a considerable amount of litter. It is estimated that up to 700,000 tonnes of used broiler litter is produced annually in England and Wales. The general preference is for wood shavings but some sawdust may be incorporated; straw, usually chopped, may also be used.

PIGS

Most fattening pigs are kept on slurry systems with the exception of perhaps the smaller herds. The great merit of straw bedding is that it provides occupation for a pen of animals which would otherwise perhaps resort to tail biting and such vices. The drawback for the pigman is having to manhandle into buildings, which are often small and low, the considerable numbers of bales required twice or thrice each week. Where a kennel system of housing is used in a tall building, then the straw required can be stored above the kennels and the bedding thrown down to the pigs, although this is far from easy when there is much straw in store. This system produces a very strawy muck requiring hand or tractor scraping for removal to the muck midden. The amount of bedding used is largely determined by the pigman's preferences and to a lesser extent by the effectiveness of the drainage system in keeping the pen area free of standing water or urine. This presumes the pens are protected from direct rainfall. Some pig keepers have reservations about wood shavings for bedding as pigs will sometimes eat too many and may suffer gut blockages as a result. However, for show-ring purposes, sawdust and shavings are frequently used as litter, being absorbent and showing off clean pigs to great advantage.

On some intensive farms, the farrowing house may still see some straw used for bedding the sow and piglet creep area. There is a noticeable trend now to turn to slurry systems in the farrowing house, again based on the reduced labour needed to operate the system.

Water

Adding water to manure dilutes its inherent fertiliser value and greatly increases the volume of slurry to be handled and stored, thus adding to direct costs. Moreover, the extra volume may well prove a headache, if not an acute embarrassment, where restricted areas of land for spreading are available on the farm. Few would disagree with these obvious facts, yet is is quite common to find slurries of only half of the dry matter at which the particular livestock excrete manure. Admitting that some extraneous water

will usually mix with the manure, there is still an oye.
indicating careless use of water about the farm or p
buildings resulting in slurry dilution. So what are the h.
of additional water? The three main sources are used cl
washing down water, rainwater and underground leakag

(i) *Cleaning water*

It will be convenient to deal with that used for equipment cleaning and then to mention general cleaning uses.

Clean milk production under the Milk and Dairies Regulations necessitates a fair amount of hard work and water to keep milking equipment clean. Table 7 is a guide to the quantities commonly used. The merits of hand cleaning where possible in reducing water requirements are nicely illustrated in this table, as is the range of usage. The economy of, say, only 50 litres in cleaning a smaller

TABLE 7. Approximate Water Requirements for Cleaning Milking Equipment, etc

Description		Litres of water	
Parlours:			
milking buckets & all units (4 abreast)			140
with jars, each unit—circulation cleaning	14		
—2 rinses	21	35	
without jars & pipeline plants, each unit		45–70	
additional lengths of pipe, per metre		2·5	
Parlour cleaning: *e.g.,* 12 stall herringbone.			
(a) hand scraping + low vol high/pressure hose*	245		
ditto for collecting yard (75 m^2)	125	370	
(b) power hose only			
walls & fittings, low vol/high pressure* ⎱ parlour	550		
floors, low pressure/high volume** ⎰ collecting yard	410	960	
Floor cleaning with low pressure/high volume hose per 100 m^2		750	
Automatic wetting of rotary parlour equipment			
(to aid easy cleaning after milking) per cow place per hour		30	
Bulk milk tanks:			
(a) hand cleaned		25–70	
(b) automatically cleaned—up to 2,000 l capacity		90–140	
over 2,000 l capacity		160–230	

* low vol/high pressure hose flow typically 0·5 l/s at 1,000 kPa
** low pressure/high vol hose flow typically 3·1 l/s at 100 kPa

TABLE 8. Water Requirements for Additional Tasks in Milk Production

Description	Litres of water
Udder washing, per cow per day	2
General washing down in milking premises:	
with some hand cleaning, per cow per day	10
hosepipe only	25
Water curtain against flies entering parlour, per hour	120

bulk tank may not look impressive until it is recollected that this is a daily saving which over the year may add up to 18 cubic metres. It represents a saving not only on the storage/disposal aspects, but also on the costs of the original clean water. Whilst touching milk production, there are additional tasks requiring water as in Table 8.

When cleaning manure from parlours and collecting yards, very little water is needed if the cowpats are shovelled up first of all. It is still a fairly common fallacy to then attempt using a mains-supplied hose for cleaning the floor. Although it may have quite a reasonable pressure, it will deliver only 20–30 litres/minute, which is quite inadequate. A better solution is a trough or troughs strategically sited and to use a hand bucket for cleaning.

If a hose it must be, then a 38 mm fixed supply pipe fitted with several quick release couplings will allow the use of a 9 m length of 25 mm hose. Supplied with water at 100 kPa (15 psi), this will deliver an adequate 200 litres/minute through a 22 mm diameter nozzle. Modern nozzles incorporating an on/off control as well as a variable spray pattern are well worth using, as it is more likely the operator will switch off when water is not needed for a few moments rather than simply throw down the hose if the tap is too far away to reach conveniently.

Other farm-cleaning operations may use water, but the drainings will not normally be directed into the slurry system. The possible exception is when cleaning out pig pens between batches. The lowest volume is needed if a steam cleaner is used and then, in ascending order of consumption, a powered hydraulic jet washer and then an ordinary mains hose. The volume used will depend on the types of wall and other surfaces being cleaned; the frequency with which cleaning is carried out; whether one man alone cleans or whether a small gang is at work and finally the standards of cleanliness desired. Estimates from operators seem to vary between 100 and 600 litres per pen.

(ii) *Rainwater*

This is perhaps the largest source of additional water in slurry systems and usually unrecognised as such. Of course most folk have a good idea of the rainfall in their area, sometimes even of the very local variations caused by nearby hills or valleys. If a farm has an average annual rainfall of, say, 600 mm, then any open storage vessel for slurry will theoretically receive that depth of water on top of whatever slurry the livestock may contribute. A statement of the obvious surely? Yet how rare is it seen to be included in the calculations of storage capacity required.

The net actual extra storage volume for rainfall will be something less than the rain occurring during the months of storage as there may be some evaporation. This will be little enough over the winter, and it is worth remembering that the rate of evaporation from the surface of a crusted-over slurry store will be some way short of the evaporation from a free water surface, upon which basis evaporation rates are customarily worked out by meteorologists. However, the really large volumes of rainfall are collected by buildings and concreted yard areas. It is an obvious golden rule that roofwater from buildings should be led away to soakaways quite separate from foul water drainage coming from stockyards, feeding, loafing and collecting areas. The yard space allowed for livestock should be minimised since this drainage will be too foul for ditch disposal and so must either be collected prior to pumping onto land or accommodated with the slurry—which looks to be convenient and is certainly a widespread practice.

A little arithmetic might be useful here to illustrate the problem. A 100-cow herd in cubicles may occupy a building roughly $31 \text{ m} \times 13 \text{ m}$ or 403 m^2. If annual rainfall is 600 mm, then the indicated total of roofwater annually is 242 m^3 and it would be folly to mix this with the manure system. Those who visit many farms will no doubt confirm how common it is to see during rain, blocked gutters or broken downpipes directing a minor torrent of roofwater into yards and so often thereby into the slurry system.

On a livestock farm, it is sometimes very instructive to work out the catchment area of yards and roadways used by animals, adding in feeding and silage clamp areas. This is a good way to concentrate the mind on how to minimise the rain-collecting area. A saving of a quarter or a third can often be made: a case may be recalled where the reduction was from 0·4 ha to 0·25. Expressed as part of a hectare, the saving looks small but it still amounted to 1,500 m^2.

METEOROLOGICAL OFFICE DATA

The Meteorological Office has now loaned a small number of staff to MAFF and the agricultural meteorologists are proving to be a most useful addition to ADAS activities. Through their channels of communication with the Met. Office, it is possible to obtain for any geographic position in UK a set of rainfall charts which can be of great interest and most helpful in planning waste handling systems. One chart displays the amount of rain falling in a number of periods, in length varying from one minute to twenty-five days. The same chart also shows for these periods, the maximum rainfall which will occur twice in one year, once in two, five and ten years, and so on.

Thus where temporary storage capacity for two to three weeks is being planned, the likely worst rainfall in five years can be read off, and knowing the catchment area, the storage can be calculated. Where transfer pumping from a collection point to the storage vessel is being planned, the pumping performance can be determined using another Meteorological Office chart which shows on a similar frequency basis the maximum rate at which rain may fall over the same given periods from one minute to several days. It is also possible to obtain this sort of information for the individual months of the year so that the fluctuations in rainfall can be taken into account in planning the storage capacity and arrangements.

At first reading, this rainfall data may well seem abstruse but its importance is considerable in that much greater accuracy is possible in planning to deal with rainfall run-off from yards, etc,

TABLE 9. Rainfall in Millimetres for a Range of Durations and Return Periods for a Location in Essex (Annual Rainfall 623 mm)

Duration	Return period*			
	1 yr	2 yr	5 yr	10 yr
0·5 h	10·5	12·6	16·7	19·7
1 h*	13·0	15·6	20·3	24·1
12 h	25·2	29·5	36·3	42·5
24 h	29·8	34·9	42·3	49·2
2 days	35·3	41·2	49·2	56·8
8 days	59·7	67·5	78·2	88·0

Source: Meteorological Office data. Crown Copyright.

* Return period: for example, once in five years rainfall during 1 h will reach 20·3 mm.

than simply averaging the years rainfall over 365 days. To illustrate this Table 9 shows data for a farm in Essex having an annual rainfall of 623 mm. The table has been extracted from the full data available, which normally contains ten columns and eighteen periods of time from one minute to twenty-five days.

(iii) *Underground seepage*

Most slurry and manure-handling collection tanks or pits and transport channels are underground. This leaves them vulnerable to the ingress of underground water, although this is unlikely on sandy soils and perhaps chalks. On most other soils, water tables rise in winter; and on wet sites the water table may be very high for long periods, sometimes permanently. The sure way to waterproof underground works is to enclose them in a metal tank and sometimes a sandwich of concrete, tanking and concrete is used. This is very expensive and not normally economic on a farm. Glass fibre and glass-reinforced-concrete preformed sections have more recently come into use successfully, but again costs are higher than for the usual concrete block and cement-rendered constructions. The latter serve quite well in many farm situations, but the claim to be able to waterproof underground channels with a polythene layer in the concrete or by using additive chemicals in the mix very often fails. The highest standards of workmanship are necessary too and these are not always employed.

The rate of in-leakage is mainly determined by the head of the ground-water over the liquid in the channel or pipe, given of course that there is a crack or other defect to let in the water. It is not always obvious either that leakage is occurring, especially if the inflow is not great. The tell-tale is when, say, a four-month storage vessel fills up in less than three months or even quicker.

The foregoing takes it for granted that such leakage is a bad thing because of the extra costs incurred in disposing of the unexpected extra volume of slurry. It is worth mentioning in passing, that an underground leak outwards is potentially an equal evil since livestock effluent may then be uncontrollably polluting underground water supplies. Under the proposed legislation of the COPA 1974, it is likely that the local water authority may be given powers to require prevention or rectification of such leaks. This could be a laborious and time-consuming task and therefore expensive.

FARMYARD MANURE

Much of the foregoing information is relevant to FYM production but there are some comments to be made about the material itself. The term is really a generalised description of any mixture of faeces, urine and litter—bedding, waste food or deliberately added absorbent material to increase the dry matter to ensure satisfactory handling during storage and transport. As a commodity, FYM tends to be more expensive to handle unless the buildings in which it is produced are readily accessible by large capacity mechanical handling equipment. One man can then handle large tonnages in a day cheaply—ignoring for the moment the problem of the capital costs of the equipment.

The density of FYM is high when in situ in livestock yards because of compaction but upon handling out into trailers, the material fluffs up considerably. In a series of tests of FYM trailer spreaders, NIAE reported densities varying from 390 kg/m^3 to 1,065 kg/m^3. A heap of FYM will often be seen to be gently 'steaming' and giving off heat. This is a sign of aerobic digestion which is made possible by the straw or other litter allowing air to penetrate to most if not all of the heap. There is then sufficient bacteria and the right conditions of air and moisture to decompose large amounts of organic matter. There is a penalty to be paid for well rotted manure and that is the greater loss of nitrogen than when FYM is compacted tightly and retained in an anaerobic state.

The smell emitted may be far from unpleasant and is often viewed by the urban public as a 'good old country smell' with connotations of 'natural processes' so beloved by the modern environmental enthusiast. When spread on the land, well rotted FYM is seldom the source of complaint about foul smells. It is this attribute of not smelling very much and being fairly acceptable to the general public which is a most important consideration not to be overlooked.

This is not to say that a very wet heap of FYM, especially from pigs, may not produce pungent and offensive smells when disturbed, so care in the siting and management of FYM stores is needed. The most important design aspects are to ensure good drainage from the midden; to provide easy access for tractors and handling equipment; and to provide adequate hard standing for the heap.

PLANT NUTRIENTS IN MANURE AND SLURRIES

Over the years, as the result of experience and the analysis no doubt of many, many samples of manure, there is now a generally accepted norm for the content of plant nutrients. These levels are contained in Table 10 for a selection of animals. It cannot be too highly stressed that the figures for slurry are for *fresh undiluted* material.

There are a number of factors which cause the actual plant nutrient content of manures to differ from farm to farm. In addition, not all the nutrients shown by laboratory analysis to be present are available to plants completely. It is easy to forget that in utilising manures and slurries, the farmer faces a difficulty in that the actual crop response he may perceive can vary because of the permutations and combinations of availability of the nutrients and the factors influencing nutrient content. If this state of affairs is appreciated, there will be a better understanding of the limitations in accuracy of planning when incorporating organic manure into the farm fertiliser policy. Perhaps that's another way of saying results can vary when using organic manures!

So what are these so-far-unnamed factors which modify the nutrient content of manures? Once listed, they look pretty obvious yet so often their reality is overlooked. They are:

TABLE 10. Nutrient Content of FYM and Fresh Undiluted Slurry (Fresh Weight Basis)

Description	% N	% P_2O_5	% K_2O	% dry matter
SLURRY				
Cattle	0.5	0.2	0.5	10–12
Pigs fed:				
dry meal & water	0.6	0.4	0.3	10
pipeline	0.5	0.2	0.2	6–10
whey	0.3	0.2	0.2	2–4
Poultry:				
hens	1.4	1.1	0.6	25
ducks	0.5	0.5	0.1	12
FYM				
Cattle	0.6	0.3	0.7	25
Pigs	0.6	0.6	0.4	25
Poultry:				
deep litter	1.7	1.8	1.8	10
broiler litter	2.4	2.2	1.4	10
in-house air dried	2.2	2.8	1.9	10

- *The type of livestock:* the amount of manure varies with the size of the animal and different species produce variations in quality of excreta: Production animals, *i.e.* animals producing milk, meat, eggs, etc, especially in intensive systems, will usually be on a higher plane of nutrition and thus tend to void more nutrients. Ruminants might be expected to excrete different amounts from monogastric animals like the pig and hen.
- *The animal's diet:* as mentioned above, the diet will be determined by the animal's function, whether growing, being carried through a store or rest period or in full production. Within this framework, any given total nutrition requirement can be met by combinations of different foods, the digestibility of which can vary, in turn, influencing the quality of the dung.
- *The conditions under which the manure is produced:* how much dilution of slurry has occurred with rainwater and wash water? Has FYM been kept dry, has it been well compacted and how much litter does it contain?
- *The length of storage and conditions in the store*, which leads on to considering:

NUTRIENT LOSSES DURING STORAGE

The story is different for FYM and slurry and so these are treated separately.

FYM Losses

There are three principal routes by which plant nutrients are lost from FYM, namely by leaching, loss of gases or seepage of liquid.

Most FYM heaps are out in the open and rainfall will dissolve soluble nutrients present in the heap. If enough rain falls to wash out the solution, the nutrients will be lost into the ground. Clearly, the amount of such loss varies with rainfall, exposure to wind (which may either drive in rain or conversely assist in drying out the heap so diminishing leaching loss), the storage arrangements (a shallow straggling heap is more vulnerable than a deep stack) and length of time in store.

Not unexpectedly, the amount of the main nutrients lost in this way also varies. Dutch investigations some years ago into this admittedly difficult problem put N losses at about 20 per cent:

phosphate at 5 per cent or just over, and potash at 35 per cent over the winter storage period.

Gaseous loss. About 10 per cent of nitrogen is lost in this way either as ammonia produced by partial decomposition or as nitrogen gas, where decomposition has progressed further to result in denitrification to nitrogen gas. Manure in a compacted heap loses little gas, whereas a loose heap, moved frequently or added to progressively, can lose three or four times as much.

Seepage. This term is used to describe the loss of liquids from within the heap—the 'gravy' in popular parlance—into the ground. In practice it would be difficult to differentiate between leaching loss and seepage since a heap of FYM usually looks to be gently weeping from its base anyway.

These losses cannot be prevented in reality, but theory suggests they can be minimised by:

storing the manure on a concrete pad;
making a deep compacted heap;
disturbing it as little as possible;
covering it to keep off rain.

Perhaps this is an opportune moment to deal with the frequent enquiry 'is it worth covering a manure heap?' Using the November 1979 costs of artificial fertilisers as the basis of calculation and the previously mentioned rates of loss, then by covering a heap a maximum saving of about 70 pence per m^3 is indicated. This rules out on costs the provision of a permanent roof or dutch barn type of structure which together with a concrete floor costs around £27/m^2. A net two metres depth of storage is the most that could be expected in practice and so a saving of only £1·40p/m^2 would not cover the amortisation cost (£5·54p/m^2) of a barn. A cheaper cover, such as a sheet, could be afforded but the physical task of dragging it into position would not be popular and there would be the problem of anchoring it against the wind. The labour cost would probably be greater than the fertiliser value saved, and subjected to such corrosive conditions and frequent handling, any sheeting would have a very short life.

Losses from Slurry

In the interests of avoiding potential water pollution, on many farm sites slurry ought to be stored in a watertight store whether it be a purpose-built above-ground store or a lined earth hole in the ground. Where seepage from the store is impossible, then it is generally accepted that no measurable loss of phosphate or potassium occurs. Nitrogen is lost as a gas, the amount depending on:

- the length of storage period;
- the type of material put into store, in turn affected by the livestock concerned and their plane of feeding;
- the weather; mainly a temperature effect, less loss in cold weather;
- the exposure of the store, which is linked to the weather effect and the amount of agitation given to the store.

It is a necessary practice to keep the store contents mixed in above-ground stores and this may be achieved by hydraulic circulation, mechanical mixing or blowing air into the store. Any form of agitation will release trapped gas bubbles and further stimulate decomposition of the manure. Thus there is a clash here between conserving nitrogen and maintaining the handling consistency of the slurry.

A number of researchers have tried to investigate the amounts of nitrogen loss in practically sized stores with very varying results. Reports are available showing losses from as little as 5 per cent to over 65 per cent; at least one result showed no loss and another appeared to have gained nitrogen over the winter storage period.

At such an illogical result and at the wide range of experimental findings, it would be natural to criticise the researchers as an incompetent bunch. In their defence, it must be said that the factors mentioned at the beginning of this section on losses from slurry will have been different during the various experiments. In addition, there remains the very perplexing problem of how to sample a large volume of slurry accurately. The laboratory analytical process may use only a few cubic centimetres and this may have to represent the bulk of the store which may amount to hundreds of cubic metres. This means the sample may be ten million times smaller than the store. Slurry stores tend to separate into a floating crust, a middle liquor layer and a bottom sludge, and so the problem is which bit of the store to sample to represent the whole volume.

This very same problem faces the practising farmer who wants to determine the fertiliser value of his store before calculating how much to spread per hectare.

It is possible to conclude that losses take place from all forms of stored manure. These losses are influenced by many factors and will vary enormously. In practice there is probably not a great deal that can be economically done to prevent losses, although it is perhaps helpful to realise how they are influenced since opportunities to reduce loss can then be recognised. In slurry storage and handling, the priority is usually to keep the store mixed and in a state to facilitate unloading and spreading to land. These considerations and others such as smell will often override the farmer's concern about loss of only some of the nitrogen value.

AVAILABILITY OF NUTRIENTS IN MANURES

Having earlier outlined the many factors which influence the manurial content of the faeces and urine excreted by livestock and then mentioned the several ways in which plant nutrient content can be eroded by losses, the manure story is not yet complete. The remaining aspect is that of availability, *i.e.* whatever level of nutrient content remains in the manure or slurry at the time of spreading, how much is actually used by the receiving crop? From the farming point of view, this is the crux of the whole business of recycling the plant nutrients in livestock wastes rather than dumping them, with the consequent risk of polluting the environment.

Before delving into details, it probably lessens the complication to say that the main concern is about the N content. Broadly speaking, P and K are reckoned to be retained and become available to crops over a longish period of time but not all in the first season of application. A very small amount of P from some soils can be washed into drainage water, but this tends to be ignored in the UK though not in Ireland where phosphate agglomeration in Lough Neah is a source of concern. Furthermore, on many soils, the P and K status is adequate and so not a factor limiting crop growth. In contrast, little N is carried over from one year to another, and so at the beginning of a cropping year N is in short supply. Hence the importance not only of knowing how much N is needed by the particular crop, but also of the amount of N which can be supplied in the season of application by the organic manure. It is by this amount that the thinking farmer will reduce his artificial N application to save money and also avoid using excessive N on the crop.

Many experiments have been carried out to determine nutrient availability and investigations continue. Not surprisingly, a range of results have been produced caused no doubt by variations in soil type and weather. Table 11 lists some commonly accepted

TABLE 11. Availability of Nutrients in Manure Applied in Spring

Description	% available		
	N	P_2O_5	K_2O
SLURRY			
Cattle	50	50	90
Pig	65	50	90
Poultry	65	50	90
FYM			
Cattle	25	60	60
Pig	25	60	60
Poultry litters	60	60	75

standards of availability, and as before, it is worth remembering that the percentages quoted are indicators and not necessarily precise levels of performance to be expected. In addition, applications at other times of the year will give usually poorer results. This is because in springtime, soil temperatures begin to rise and with them the whole pace of soil-borne activity including the most intense rate of plant growth—and therefore of food needs—during the year. When manure is added to this dynamic microcosm, the plant nutrients are quickly anchored in the soil and worked upon thereafter by soil fauna to make them available to plants. As a result, in a normal spring little of the manure is wasted.

In contrast, when manure is applied in autumn, the whole of the winter rains are available to leach out any soluble fractions into the drains leaving little for the next year's crop. Manure applied in summer can be very variable in effect since so much depends upon there being enough rain to wash the manure into the soil satisfactorily. If not, then large losses of N occur from the surface of the field to the air or else the fertilising value does not reach down to active root level because of dryness. These seasonal effects are summarised in Table 12.

By combining the information in Tables 11 and 12 it is possible to draw up a further table (13) showing the available nutrients in manures and it is these quantities which can be used for planning the farm fertiliser policy. The remaining caution is that the time of application factors in Table 12 should be applied to the N totals only in Table 13 if the manure or slurry is spread at times other than in spring.

TABLE 12. Time of Application of Manures and Availability of N

Season	% of N available for crop
Autumn	0–20
Early winter	30–50
Late winter	60–90
Spring	90–100
Summer	weather dependent

TABLE 13. Available Nutrients in Manure Applied in Spring

Description	N	P_2O_5 kg/m^3	K_2O
SLURRY, undiluted			
Cattle	2·5	1·0	4·5
Pig	4·0	2·0	2·7
Poultry	9·1	5·5	5·4
FYM		kg/tonne	
Cattle	1·5	2·0	4·0
Pig	1·5	4·0	2·5
Poultry deep litter	10·0	11·0	10·0
Poultry broiler litter	14·5	13·0	10·5
Poultry air dried in house	25·0	17·0	14·0

Slurry diluted 1:1 with water divide figures by 2
Slurry diluted 1:2 with water divide figures by 3

FINANCIAL VALUE OF MANURES

Having followed the trail of excreta from the animal, through storage to the application to crops, there has been frequent mention of the variability of organic manures and of the likely losses of nutrients. However, it is not all bad news because whatever the scale of loss, nutrients which remain available to crops in the season of application have a definite financial value since they can substitute for bought-in artificial fertilisers.

The financial value can be calculated using data in Table 13 once the cost of bought-in fertiliser is known—the prices of straight N, P and K fertilisers are used. It is apparent that the price of fertiliser varies almost farm to farm, with bulk quantity and out-of-season purchase discounts available, let alone the source of purchase and the overall trading volume with the particular merchant. However, as an example, with N costing 27·6 p/kg, P_2O_5 24·1p/kg and K_2O

TABLE 14. Equivalent Financial Value of Fresh Undiluted Slurry
(based on artificial fertiliser prices, February 1980)

	$1\ m^3$ (1 tonne)	Value of slurry (£) Daily production from:
COW		100 cows*
N	0·75	3·08
P_2O_5	0·24	0·98
K_2O	0·62	2·54
	1·61	6·60
PIG		1000 pigs**
N	1·20	4·80
P_2O_5	0·49	1·96
K_2O	0·37	1·48
	2·06	8·24
POULTRY		1000 laying hens
N	2·73	0·31
P_2O_5	1·34	0·15
K_2O	0·74	0·08
	4·81	0·54

* Total slurry output *e.g.* winter-housed cows.
** Dry meal feeding.

12·7p/kg, current in November 1979, the equivalent value of manures is as shown in Table 14. The values shown are for the estimated *available* nutrient content of *fresh undiluted* manures and so represent a maximum value. The actual value will be reduced by the losses experienced in storage, etc, as outlined previously, and this will vary depending upon the individual farm practices, climate, storage period, etc.

The greatest value generally lies in the N fraction, which sadly also suffers the storage and other losses. Table 14 also illustrates the relative values of the principal nutrients and the opportunity has also been taken to include the financial values of a few convenient volumes of slurry. The table has been restricted to slurry, partly in the interests of brevity but mainly as this tends to be the most popular topic with farmers who enquire about the value of manure. Table 14 uses as its base the costs of fertiliser at 1 February 1980 and these will become out of date fairly rapidly.

Accordingly, set out below is the arithmetic in recalculating the values with increasing fertiliser costs.

Suppose that unit costs of straight fertiliser become N 30p/kg, P 28p/kg and K_2O 14p/kg. Then the value of 1 tonne of cattle FYM would be:

Available nutrients (Table 13) in kg/tonne × new price p/kg

i.e.
1.5 N kg × 30p = 45p.
2.0 P_2O_5 kg × 28p = 56p.
4.0 K_2O kg × 14p = 56p.

£1·57p/tonne

REFERENCES

Planning for Parlour Milking. *Management Aids*, **16** ADAS.

PLATE 1
Lagoon storage showing sludge deposition and crust formation in primary compartment and clearer liquor in secondary basin. _{British Farmer & Stockbreeder}

PLATE 2
Above-ground manure storage system by Farm Storage Ltd, incorporating walls designed to allow liquid escape. _{Farm Storage}

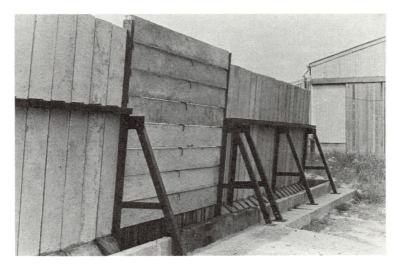

PLATE 3
Wall bracing and seepage collection gutter of Farm Storage system.

Farm Storage

PLATE 4
Concrete panel store by Reco Ltd. Tractor-driven slurry pump is filling tanker and is also connected to a jetting kit on the top edge of the store. The large diameter return pipe for unloading the store is also shown.

PLATE 5
Careful siting can blend a large slurry store into the scenery.

Howard Harvestore Ltd

PLATE 6
Slurry store agitation by recirculating slurry under pressure can be effective. A multiple outlet kit in which each outlet can be controlled and rotated 180° horizontally.

Howard Harvestore Ltd

PLATE 7
Slurry store agitation by an air bubbling system, showing the air distribution pipework secured to the floor of the store.   Howard Harvestore Ltd

PLATE 8
Slurry store agitation showing position (on brick plinth) of 3 kW compressor and piping. Also shown is the larger diameter slurry loading pipe from the below ground pneumatic pump.   Howard Harvestore Ltd

Chapter 5
STORAGE

To some it is perhaps debatable logic to move from considering the quality of manures and their production to the question of their storage without first having considered the problem of handling the various materials. At the practical farm-end of the business, the usual route is to identify the quantities of the various wastes to be accommodated and then to ponder the problem of storage in conjunction with the limitations imposed by type of farming, soil type, climate, etc or the ease or difficulty with which the manures can be disposed of onto the land.

Having sorted out priorities here, the decision-making process can move on much more informed to consider the type and housing and from that the sort of handling system and equipment necessary to move manures from housing to storage. This may well reek of heresy to the keen stock farmer who will naturally tend to choose the housing system first and reckon to be able to buy enough tackle to shift the muck as required. Perfectly possible of course, but times change and the pressure on agriculture not to pollute the environment grows, not least with some of the recommendations made by the RCEP on agriculture whose report was published on 18th September 1979. Thus it is important to identify any constraints on a new or enlarged livestock unit from the waste disposal angle before committing funds, possibly irrevocably, on a housing system which produces either more manure than the farm can deal with or manure in a form which is an embarrassment to the farm and its circumstances.

To illustrate this, examples could include expanding a pig herd when the farm is already on the small side in acreage or choosing a slurry system for the extra pigs when the existing herd production of liquid muck provokes complaints from neighbours during handling and spreading. Naturally enough, if an existing livestock enterprise is based successfully on a particular form of housing, there is a strong compunction to use the same housing for any proposed expansion. Even so, to look very carefully at the manure storage and disposal system for the expanded enterprise is sound common sense.

It is not the purpose here to become embroiled in the constructional details of various storage and other structures as this is the province of the architect and designer. Rather, the hope is to raise some of the problems, solutions, limitations and difficulties which are worthy of care in making the best choice of storage system. In addition, some of the major guidelines in successfully managing manure stores can usefully be touched upon.

WHY STORE?

Immediately there is a basic question to be faced: 'Why store?' since it costs money and needs management effort. In the days of 20- and 30-cow herds, the amount of excreta to be cleared up and disposed of was not great and on soils suitable for trailers throughout winter, very successful systems have been operated whereby each day the muck was put into a trailer, taken to the field and spread—simple, easy to manage. However, once larger numbers of cows are milked, then the daily quantity exceeds one trailer load and the time spent collecting, carting and spreading muck begins to get out of hand. The number of farms able to accommodate a daily spreading operation is reducing if only on account of the labour requirement and its disruptive effect on the daily farm operations.

Another cogent reason for storing is that with the greater efficiency of modern farming, even where soils are drained enough or sufficiently light to allow traffic in wet weather, there is often just not land free and available to receive muck because of more intensive cropping.

Probably to the manager of a farm, the greatest reason for storing manure is that it greatly reduces the daily management burden. The problem of muck carting and spreading can be put off to only one or two occasions in the year which often can be planned ahead, thus minimising headaches. This releases the manager and his stockman to spend more time tending their stock and that is where the profit is made.

Mention of profit is a reminder that one of the great attributes of storage is that the farmer can plan to recycle the manurial nutrients to crops by choosing the time of spreading and so extracting best value from the manure. In addition, by pursuing an intentional policy of utilisation, the danger of over-applying manure—which so often occurred when 'disposing' of muck on sacrifice areas—is eliminated. As has been stressed continuously, there is a most pressing need to avoid pollution.

A remaining and very important reason for storing manure is to reduce pathogens. Many animals have a sub-clinical level of disease, *i.e.*, they harbour harmful bacteria but at a level too low to show obvious signs of the particular disease. They themselves have come to terms with the infection but remain carriers of it and so excrete some of the disease organisms. If such infected muck is spread on grazing land, there is a real danger that other animals may pick up the infection. Not every herd excretes harmful pathogens but probably more do so than the farming community realises. In one survey 11 per cent of cows and 22 per cent of pigs were infected with salmonella. Although there are a number of diseases which may be involved, most can be reduced in virility or even eliminated by storing manure. The heat of the FYM heap and the conditions in a slurry store take their toll of disease.

Thus the answer to the question 'Why store?' can be summarised as:
1. Daily spreading is becoming more impractical with larger herds, less labour and tighter cropping timetables.
2. Reduced stress on daily management of the farm.
3. More time available for livestock tending.
4. Better utilisation of nutrients in manure.
5. Less likelihood of pollution.
6. Pathogen control.

STORAGE OF FYM

(a) To deal shortly with temporary field heaps first of all—there are several major considerations in choosing a site:

- The heap should be kept well away from ditches and other water courses so that 'gravy' cannot reach and pollute them.
- Fly and smell nuisance complaints may arise if the heap is near domestic dwellings.
- The site should offer a firm base which will withstand the chewing up tractor fork lifts may cause when removing the heap.
- Where a heap is gradually formed over the winter period, a good all-weather road to the site is desirable.
- The site chosen, in relation to the farm's layout, should be the best compromise between journey time during building the heap and when spreading on the land later.

STORAGE

- It seems pretty obvious to keep well away from overhead power lines, especially in view of the height to which some tractor-mounted loading equipment can reach.

At the time of writing, the implementation of the Control of Pollution Act 1974 Part 2 has been delayed again. As the Act is written, it is possible that a licence may be needed from the local water authority for even temporary sites, but the position is yet to be clarified.

The best pathogen control is likely when the heap has least surface area, *i.e.*, a cube rather than spread out as a rambling chaos. Building a deep heap is not easy and some sort of supporting sides are helpful. Portable silage clamp sides or a row of big bales can be used presuming that no one has the time nor inclination to hand build the sides of a stack nowadays.

(b) Choosing a permanent site for a muck heap or midden deserves the most careful thought, especially as it is commonly placed amongst the farm buildings.

- Access for equipment during loading and particularly unloading is most important. It is sometimes forgotten that access is usually required for the tractor fork lift and the tractor and trailer being loaded. Where high rates of work are planned, then more than one tractor and trailer may need to be in the area at the same time if the loading operation is to proceed uninterrupted.

- The site will need to be concreted as there is not really a rival flooring material. Relying on a hardcore or railway ballast bottom is a short-sighted policy. The material is removed from the store during unloading and may cause damage to the spreaders. Worst of all is the damage which may be done to field equipment during harvesting, after spreading on, say, grassland.

- The midden area needs to be laid to a planned fall to drain off seepage. This cannot be disposed of into a ditch because of its polluting nature—typical levels might be 2,000 mg/l BOD and up to 6,000 mg/l SS. The answer is to lead the seepage into a below-ground collection tank from whence the 'gravy' can be spread onto land either by a low rate irrigation pump system or more simply by periodic emptying using a slurry tanker.

  Since this tank will also receive rainfall on the midden, sizing the tank will require careful thought, about which more later. The below-ground tank should not be fitted with an overflow.

- The midden should be provided with sides to allow a deep heap to be built. Some sections of the sides may need to be demountable to allow access at unloading time. It should be decided whether the sides will be solid or provided with drainage holes in, say, the bottom metre of the wall. If drainage is intended through the walls, then there will need to be a 300 mm wide by 250 mm deep collection channel sloped 1 in 60 to 1 in 80 around the perimeter to duct seepage to the collection sump.

The height of the sides need not be unduly high as there are not many tractor loaders which can stack at great depths. Walls 2 m or 2·5 m high are probably adequate; higher than this the structural requirements of stanchions increase markedly, as does the cost.

The advantages of maintaining a deep heap are better pathogen control and less surface area for the release of odours although flies may sometimes be the greater headache, in which case the smallest area to be sprayed the better.

- If a sloping site is available, advantage may be taken of this to ease loading or drainage problems. A convenient layout is walls on three sides and the fourth side open. Trailer loads can be dumped for several days or longer before the tractor front fork need be used to pile up the manure.

There are divided opinions where a sloping floor is used. Running down into the midden ensures that the maximum amount of liquid is absorbed by the straw or other litter as well as making certain that all liquid runs into the bottom of the heap. However, even with a perforated wall behind, this tends to make a very wet heap which is undesirable. A gentle slope (1 in 100) outwards, *i.e.*, from back wall to open front, will encourage liquor to leave the heap and be collected in a wide, shallow duct running across the open edge of the midden and leading to the collection sump.

- A special note here about poultry droppings. If neat or mixed with only a little litter, then droppings are low in dry-matter content—25 per cent—and so tend to be a semi-solid which will slump and cannot be stacked. If this then gets rained upon in the midden, what was formerly an almost stable shallow lump can become a flowing mass and as such may reach the drainage system and fill it up. Where wet poultry manure is to be stored and the midden is sloped down into the back wall to hold the muck in position, the retaining wall will be subjected to as much thrust as if it were a dam wall and must be designed as such.

- On page 55 the costs of roofing a midden were said to be uneconomical, considering the question of eliminating the loss of nutrients by leaching during storage and applying the usual investment amortisation at current rates of interest.

The calculation is set out more fully as follows:

Total nutrients in one tonne of FYM (Table 10) in kg: 6 kg N; 3 kg P_2O_5; 7 kg K_2O.
Assumed rate of loss: 20% N; 7% P_2O_5; 35% K_2O.
Value of equivalent straight fertiliser: 27·6p N; 24·1p P_2O_5; 12·7p K_2O.

Therefore, if this leaching loss is avoided, theoretical financial saving achieved is:

$$20\% \times 6 \times 27\cdot6p = 33p$$
$$7\% \times 3 \times 24\cdot1p = 5p$$
$$35\% \times 7 \times 12\cdot7p = 31p$$

Total saving = 69p per m^3

Dutch barn 6 m to eaves and 13·7 m span, including concrete floor approximately £27/m^2.
Amortised over 20 years at an interest rate of 20%, annual charge £205 per £1,000 capital cost, or £5·54p per m^2 of barn floor.
At even two metres depth of storage this accommodates only two cubic metres of manure, or a saving of $2 \times £0\cdot69 = £1\cdot38p$.

Sizing a FYM midden

The volume to be stored will be the total of daily excreta and bedding × number of animals × length of storage period in days. For example, consider a 100-cow herd littered with enough straw to produce FYM so that all manure can be solid handled. The theoretical volume of storage required would be calculated as follows:

excreta/cow/day = 41 litres = 0·041 m^3
straw litter (density 100 kg/m^3) 3·5 kg/cow/day.
 volume of straw/day $3\cdot5 \div 100$ m^3 = 0·035 m^3
i.e. (0·041 m^3 excreta/cow/day + 0·035 m^3 straw) × 100 cows × 180 days = 1,368 m^3.

However, the actual volume required would be rather less because of the absorptive capacity of the straw (about 2·5 times its own weight of liquid) and also the very variable amount of compaction achieved in the heap. Usually, there will be some guide on the particular farm as to the amount of litter used and the yard space occupied with FYM from the livestock enterprise.

This may appear an imperfect answer to a simple problem but the volume of FYM is so much influenced by the amount of straw added. If a midden becomes too small for the winter's storage, it is usually possible to move some of the FYM to a temporary field heap, especially when the ground is frozen. Many farmers do this as a matter of course and so economise on the size of midden area needed.

Sizing a tank to collect seepage from a midden of FYM

Since by definition FYM is a solid material, then the tank has to cater for 'gravy' exuding from the heap as the result of rainfall. On many sites in the UK, approximately half of the rain falls in winter, and so it would be a possible method to halve the annual rainfall and divide this by the six months of storage to give the indicated monthly capacity of the collection tank—presuming that emptying the tank once per month is acceptable. However, much greater accuracy is obtained by using the rainfall shower frequency tables mentioned earlier on page 50, or referring to MAFF Technical Bulletin 35, *The Agricultural Climate of England and Wales*.

Although weather data shows there to be some evaporation of moisture during the winter period, this is best ignored since the tank must cater for a month's rain in the wettest part of the year when there is likely to be little evaporation. Thus the calculation is to determine the area of the midden—the whole concreted area, not just the area upon which the FYM is heaped—and to multiply this by the rainfall in the wettest month of the winter, *e.g.*:

Midden area 10 m wide and 15 m long.
Annual rainfall 713 mm ∴ monthly average 59 mm but
Wettest month November 82 mm

∴ Indicated tank capacity

$$10 \text{ m} \times 15 \text{ m} \times 82 \text{ mm or } 10 \times 15 \times \frac{82}{1,000} \text{ m}^3 \text{ (1,000 mm in 1 metre)}$$

i.e. 12·3 m^3

Allow 10 per cent for free-board, sludge and other detritus in the tank bottom taking up storage volume,

$$i.e. \; 12 \cdot 3 \times \frac{110}{100} = 13 \cdot 53$$

∴ Plan on tank of 13·5 m^3 capacity.

STORAGE OF SLURRY

There a few topics of general relevance which might be aired conveniently here to avoid endless repetition when considering the various systems. Despite the problems slurry can bring, in the last few years the slurry system has become very widespread in the pig world, and more and more dairy herds are going over to slurry—most large herds are already practitioners.

As was mentioned earlier, slurry is not a consistent material and may contain a variety of additions. Left undisturbed for a while slurry separates out into fairly distinct layers: cow slurry into bottom sludge, a middle layer of true liquid and a floating crust of fibrous material. Left long enough, this crust can reach a metre thick and may develop a top growth of vegetation including bushes and even young saplings. It cannot again be liquified and so can present a sore problem at unloading time. This is the result of BAD MANAGEMENT.

Pig slurry will separate into a bottom sludge and a top clear but dark liquid. In warm weather small bubbles of gas continuously break surface due to some anaerobic digestion taking place. In some cases, a crust of fibre and pig hair will form on the surface but usually not for quite a long time after the store was loaded.

Poultry slurry may form something of a crust usually not very thick and resembling a slightly dried out surface.

For stores relying on liquid handling equipment, crusting must be avoided by good store management and regular recirculating of the tank contents.

Advantages of Slurry

1. The over-riding attraction seems to be the halving or greater reduction in labour needed for livestock manure handling.
2. The bedding needs are also greatly reduced if not altogether eliminated. The cost and time spent straw handling during the harvest peak are avoided.
3. A range of purpose-built equipment for handling slurry is readily available.
4. Planned properly, a slurry system can deal with all manures on the farm using the one set of equipment (pump and tanker), whereas a solid system often produces some liquid manure, thus needing solid and liquid handling equipment.

Disadvantages

1. Undoubtedly the greatest drawback is the anaerobic conditions during storage of manure. This results in the formation of complex organic compounds which when agitated or spread on the land release the most awful smells on occasion.
2. Livestock buildings designed for slurry are usually more expensive than for straw-bedded systems. The necessary storage and handling facilities are more expensive too.
3. Higher standards of daily management are necessary for success with slurry systems.
4. It is too easy to increase inadvertently the volume of slurry to be handled, by careless use of water about the farm.
5. As a commodity to work with, slurry is much more offensive than FYM.

STORAGE SYSTEMS

It is difficult if not even contentious, to try to categorise the many systems of storing slurry. Using cost per unit volume stored as a basis, then the order looks like:

lagoon; compound of all earth construction, or with concrete bottom, or of all-concrete construction; also above-ground steel and timber concrete-bottomed compounds; above-ground sprayed concrete structures; concrete block or panel tanks; above-ground steel stores. In some countries, circular above-ground stores of wood or wood and steel composite construction are available.

(1) Lagoons

(a) It is important to draw a distinction here as the word 'lagoon' is used indiscriminately. In the strict meaning of the word, reference is made to the storage in a shallow (about 1–2 m deep) pond of organic wastes. Due to the combination of anaerobic conditions at the bottom of the pond and aerobic conditions near the surface, digestion of the degradable organic matter can occur *but* this will only take place if temperatures are sufficiently high for long enough in the year. This method of 'treatment' of manure was popular in America, and some years ago the idea obtained its enthusiastic disciples here in the UK. None of the examples built proved to be other than a simple holding store. The initial claims for success were likely explained as being some evaporation from the surface plus liquid leakage out through the sides and bottom.

The amount of oxygen which is estimated to be dissolved into a lagoon is in the range of 20–50 kg per hectare per day. To satisfy the BOD requirement of a year's manure from, say, 1,000 fattening pigs, a pond of between 2 and 4·5 hectares would be required and even then the supernatant would not be fit to discharge into running water and would therefore have to be either further treated or spread to land. This probably makes clear the irrelevance of the large 'treatment' lagoon under UK conditions.

(b) The description 'lagoon' in the UK often refers to a below-ground or part below-ground excavated hole into which slurry, usually from pigs, is deposited. Other names include 'slurry bowl' or more descriptively 'a wet hole'. The main characteristic is that the store is wet and the material will flow readily enough to be handled as a liquid. Emptying is usually by slurry tanker either pump or vacuum filled. Whilst there are successful examples of such lagoons, there is a basic flaw in that they infer the handling of large quantities of water which has become mixed with the original excreta. Extra water means extra costs for handling and transporting. The extra volume may, in addition, be a headache to accommodate on the land area available for spreading.

For cows on a straw litter/slurry system, a lagoon is best avoided, as the straw forms wads which can effectively block the unloading equipment. Pig manure is easier to unload as the top liquid layer flows readily to a tanker on the bank-side. However, if no effort is made to remove the bottom sludge by stirring up the lagoon before unloading, the sludge will be left behind and after a year or two will present a sore problem of removal.

Many slurry compounds end up as lagoons because the site chosen is too wet, ground water leaks into the manure and very wet conditions result. The manure is held in an anaerobic condition with consequent development of foul odours which will be given off steadily much of the year. In summer, digestion rate increases and so complaints about smell may arise.

Looking to the future, it seems almost certain that before constructing a new lagoon the express permission of the local water authority would be needed for the site since Part 2 of the proposed COPA requires such licensing.

SITE

Situating a lagoon near any watercourse is unwise as in the event of misfortune there is little margin for error. Similarly, a lagoon on a hillside is asking for trouble.

Construction

Remembering that lagooning is storage in a liquid form, then it is obvious that the site should have a soil type which will not allow leakage into or out of the lagoon. In-leakage will mean the risk of trying to drain the local parish at unloading time, which would be a costly nonsense. Leakage from the lagoon could result in pollution of local waters and the ultimate prospect of prosecution.

Where the site is on permeable soil, then a lining of 300 mm of puddled clay can be used for waterproofing, but this is a bit theoretical since the lining may be damaged at unloading time and certainly will want periodic maintenance.

Of absolute importance is that the excavation and retaining bank construction should be properly done and be incapable of sudden failure—it is just not acceptable to run such a risk. As a general guide the slope of the internal face should be close to $1:2\frac{1}{2}$ and that of the outer face $1:2$. Depth of storage is best kept shallow, say 2 m. Because the contents of the lagoon closely resemble water, the banks should follow reservoir practices and be of adequate size and constructed properly. Perhaps the best piece of advice of all is to employ the services of an experienced civil or water engineer to investigate the individual site, take and test soil cores and to design the bank accordingly.

Making the bank is not just a matter of using a bulldozer to excavate a hole in a field and push up the earth all around the sides, as this will result in an unstable bank. For a large lagoon, the construction is best left to an experienced contractor.

Management

One of the attractions of a lagoon is that it demands little attention. However, this does not mean none at all! The most important point of all is to ensure the lagoon does not overflow at any time; thus regular inspection of level is necessary and sufficient free-board at the edge maintained to cope with sudden rainstorms. The banks should be kept in good order and any damage sustained during unloading, especially rutting of the top, should be rectified immediately, not left until 'later'.

The regular inspection mentioned above should also include a close look at the safety fencing and gates. It is most important that these are maintained effectively, and a regular check is the way to ensure this minor task is not forgotten.

Perhaps the most difficult management task to 'get right' is deciding when to empty so that the least possible nuisance from smell is created. Waiting for a favourable direction of wind, plotting sensible transport routes avoiding difficult neighbours and choosing fields away from habitation are three common aims, but in practice they are often very difficult to achieve within the time available for spreading as dictated by crop and seasonal requirements.

(2) Compounds

This is a fairly broad description of a structure into which cow slurry, with or without straw or other material, is transferred for long-term storage. The usual aim is to encourage some drying out so that the bulk can be handled as a solid or semi-solid when emptied, typically in late summer onto stubbles. The compound may be partly below and partly above ground or wholly above ground. Construction can include entirely earth-built versions; the same but with part or all of the floor in concrete; steel uprights supporting wooden walls with a concrete floor; and all concrete variants of the same idea.

Before dealing with these in turn, mention ought to be made of factors influencing the success of a compound. There are many good examples of compounds in use and they have two outstanding attractions: cheapness and convenience. Clearly, cost is least with an all-earth structure. Probably much the greatest attraction is that a well-designed compound will accept anything the tractor scraper can push into it. An odd claim perhaps, but to the working farmer it is comforting to know that the slurry system will work dependably and is not sensitive on a daily basis to the whims of mechanical reliability—always assuming there are enough tractors to keep the scraper functioning.

Lumps of unwanted bedding, silage, etc can be pushed into the compound and will pose no real problem at emptying time. This daily reliability frees the dairy man to concentrate on cowkeeping in the knowledge that three or four days' activity in summer will see the compound emptied and ready for the following winter. Such practical considerations outweigh the loss of much of the N element in the manure due partly to leaching and volatilisation during storage and partly to the losses sustained as the solid material is spread and left lying on the field exposed to the weather.

Siting

At present no official permission need be sought over siting but under proposed legislation, a site licence would need to be sought from the water authority. Just when this provision will come into force is not at present known but even so, the responsible farmer will voluntarily discuss siting with the water authority now since no one wants to pollute water supplies deliberately. Even without the onset of the proposed legislation, a compound causing pollution can land its owner in court anyway and so it seems sensible to seek water authority advice, especially if the soil is permeable.

Remembering that the aim is to encourage drying out of the contents, the site should be dry and away from water courses too. If the land slopes, this might be a minor advantage but more likely a bigger headache as it can sometimes be a problem to prevent surface rainwater getting into the compound and great care is needed to ensure a stable retaining bank. An undrained field site is less of a problem. Where under-drainage exists, seepage from the compound may enter the drainage system. To prevent this, drain runs should be diverted around the actual site and this is not always so simple.

A sensible precaution, but surprisingly overlooked sometimes, is to avoid a site beneath electricity power lines lest there should be an accidental flash-over onto any tall equipment, *e.g.*, loaders working around the compound. It goes without saying that the position of the compound should be very carefully thought out. Situated close to the housed animals will entail least work in its daily loading with the scraper. It is not often that the nearness of other buildings, roadways, etc allows a free choice of site and enough room must be left for tractor manoevering during scraping and access during loading.

Sizing

Using the information in Table 4 it is simple to calculate the cubic storage requirement for the herd size and length of housing period. Dairy washwater is better excluded from the compound, but if some water must be accommodated then its volume, and the depth of winter rains should be allowed for in the calculation of size since all of these will have to contained before warmer weather and thus evaporation in early spring come along. It's prudent to allow a small surplus so that overflowing is not a possibility.

For example: 100-cow herd in cubicles all slurry and straw bedding, 180-day winter; farm has 900 mm annual rainfall and some washwater must go into compound.

100 cows @ 41 l/day manure = 4,100 l
100 cows @ 5 kg/week bedding = 71 kg/day (absorbency 2·5 times)
∴ equivalent to 29 kg
100 cows @ 14 l/day washwater = 1,400 l
                total daily volume 5,529 l or 5·5m^3
For 180 days, $180 \times 5·5 = 990m^3$
Say, compound will operate at a net 1·8 m depth.
Then area of compound required 990 m$^3 \div$ 1·8 m = 550 m^2.

Having decided the nominal area of compound required, the length and breadth dimensions must be chosen with some care. It may be that site limitations, due to the proximity of buildings or roads, will dictate one of the dimensions.

From a practical point of view, the method of emptying is a most important consideration as this will indicate maximum dimensions for the width depending on the equipment to be used. Very often the basic decision is whether to plan to empty and spread using farm staff and equipment or to use a contractor with high capacity equipment. The latter arrangement will usually see the whole job done in 2–3 days' hectic work using many tractors and spreaders loaded by a digger-loader. Where suitable contract services are available, the system has much to commend it on the basis of both cost and speed of operation—the main slurry-handling problem is dealt with once a year in a few days leaving the farm staff free to concentrate on cowkeeping for the rest of the year.

Where the intention is to use contract services, then it is a very sound idea to discuss the layout and design of the compound with the contractor (who may be able to excavate the compound too) so that maximum lengths of reach of his equipment are taken into account. A happier contractor will probably result in quicker emptying each year subsequently.

(3) Earth-banked Compound with Earth Floor

Probably the cheapest method there is of storing slurry; the price is mainly determined by the cost of the bulldozer which in turn is cheapest where the contractor has ample warning and can fit the job into his own work schedule conveniently rather than have to undertake the task at the last minute or when ground conditions are difficult. More than likely, the necessary safety fencing will be the most expensive single item of the total cost.

Once the site is chosen, a few test holes with a spade is time well spent to determine at what depth the natural water table lies.

Remembering that the aim is to use the summertime evaporation to dry out the slurry, there is little point in digging the compound too deep and getting into the water table.

Emptying

There are three options open to emptying the compound:

(i) Use a digger loader working from the bank top, reaching down into the compound to extract the manure, swinging 180° to load into tractor-drawn trailers or spreaders running alongside but outside the compound. One such loader can adequately supply six or more spreaders (depending on the distance and the journey time to the spreading point). Compounds containing the winter's manure from 120 cows have been seen emptied in three working days including spreading onto fields.

(ii) Where a dragline excavator is available, a narrower bank can be used and the dragline operated from outside the compound altogether. This is fairly rare nowadays as the cable draglines are being superseded by modern all hydraulic designs, though of reduced reach.

(iii) Enter the compound with a tractor or other type of loader which then loads into spreaders alongside. This means spreaders must also enter the compound, which is fairly unsatisfactory because wheelgrip can be a problem, manoeuvring is difficult, time is lost, everything and everyone gets plastered and the bogging down of only one trailer can stop the whole operation. If this method must be used, then it is important that spreaders can enter and leave by separate ramps or exits cut in the bank, travelling only forwards without the need to manoeuvre. It is also very useful if each spreader is fitted at the rear with a fender of tyres so that a following tractor can give a helping push to start a loaded spreader to move forward in the compound.

That the retaining banks should be stable and secure has already been emphasised, and so adequate width is essential. The angle made by the bank face can vary between 30° and 45° with the horizontal, the steeper angle being restricted to heavier soil sites. Top width of the bank is dependant upon what, if any, vehicles may run along the top. Thus, if the unloading system will use a swinging arm digger loader working from the bank, a width of 3 m or more will be needed.

A typical cross-section is shown in fig. 5. To prevent the bank

slipping sideways, the key-trench is an essential but often omitted feature. Briefly, the sequence of construction is:
1. All organic containing soil (topsoil) is removed from the site and stored for final dressing of the finished bank to encourage grass growth.
2. The key trench is cut out.
3. The store below ground is excavated, using the spoil to fill the key trench and to build up the full bank cross-section in layers 150 mm at a time, care being taken to wheel down all of one layer before the next is added.
4. The completed bank is finished off with topsoil and seeded to grass.
5. Secure fences and gates are erected to guard the whole installation against children and others.

Floor

This should be excavated to a flat base with a positive fall along the length so that greatest depth is at the input end of the compound below the ramp up which the tractor scrapes the slurry. To most people this may seem to be the wrong way round for the slope—'Surely the slurry won't run away from the ramp to fill the compound' is their likely reaction.

As has been described, the compound will receive animal faeces, urine, rainwater plus sundry other solid material, notably straw. By having the deeper end below the ramp the liquids are retained in this area. This allows incoming loads pushed off the ramp to distribute themselves evenly in the pool of liquid slurry. In this way the compound can be evenly and progressively filled from the ramp.

If the compound floor slopes away from the ramp, the 'gravy'

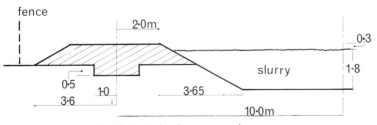

Fig. 5. Earth bank compound.

tends to run away from the fibrous material which then tends to collect under the ramp. Over a period of time this 'pit heap' of straw under the ramp can build up, obstructing the ramp and preventing the slurry from entering the compound—it tends to flow back down the ramp. By the time of the year this has happened, there is often too much slurry in the compound to allow a tractor to enter and attempt to move the heap under the ramp. The job is impossible and dangerous to do using hand labour, leaving the farm without an effective slurry system. Thus a floor slope towards the ramp is essential; 1 in 50 or 60 is reasonable. After the compound floor area has been scraped out, it is important to look for any broken land drains crossing the site and to carefully seal them outside the compound. This is to prevent 'gravy' from the compound travelling into the drainage system and so polluting watercourses, possibly at considerable distance from the site itself.

(4) Earth-banked Compound with Concrete Floor

Where compounds are emptied without entering with vehicles, long reach loaders are necessary which are usually contractor-owned. Contractor-emptying is a once-a-year event convenient and attractive for some farmers, yet others would prefer to empty on several occasions hoping to time application to a number of crops to make best use of the fertiliser value of the manure. On other farms, several sessions at intervals of unloading can perhaps be undertaken by the full-time farm staff as an off-peak job.

All of these situations are greatly facilitated if the compound has a concrete floor. Farm tractors and trailers can then enter and manoeuvre without fear of getting stuck. In addition, work rates can be quite a bit higher because of the easier travelling and better surface from which to pick up the manure.

Floor

Again, this should be laid with a slope towards the input point. Nowadays concrete is not cheap and so value for money is important. Any temptation to cheapen the quality of the floor should be resisted. The floor will have to withstand attack from the manure and wear and tear from the unloading equipment. Tractor front loaders develop high wheel contact area loadings on the floor when tearing out loads from the heap. Tractor shovels will scarify all but the highest quality surfaces, which may lead to cracking and premature failure of the floor. A badly failed floor is a great nuisance since lumps of concrete and under-lying hardcore may be dumped into spreaders, causing damage to them and possibly to harvesting machinery at work the following season in the field.

Since site conditions can vary so much, the best advice is to seek competent opinion on the depth of hardcore and concrete necessary; also on suitable additives for the mix. At least 150 mm of concrete are likely, but steel reinforcement is only needed on the more difficult sites. The work should be done with as much care as goes into the preparation of a floor in a building or yard since sloppy work will reveal itself all too soon under the corrosive conditions in a compound.

A number of compounds have been constructed with only part of the floor concreted, in the attempt to save costs. Usually a centre strip some 4–5 m wide along the length of the compound has been laid. The theory was that unloading equipment can work from this strip, reaching to the sides for complete unloading. To assist this, the earth floor sloped up gently from the concrete strip to the side wall or bank, so encouraging manure to flow down onto the strip for unloading. In practice the idea has often been less than effective. Work rate is slow as trailers must back into the compound and inevitably sometimes run off the edge of the concrete strip. Because of the wet conditions, very deep ruts alongside the strip can occur such that trailers can become stuck and damaged, especially their tyres, on the edge of the strip. In some cases, two or three years' operations resulted in undermining of the concrete edge and its subsequent failure by cracking.

(5) Concrete Bottom with Sides Designed to Allow Liquid Escape

There are two types of these above-ground structures:

(i) The traditional design was to use vertical rolled steel joists or channels at suitable intervals to carry ex-railway sleepers horizontally, which were simply dropped down the channels to form a thrust-resistant wall some 2 to 2·4 m high. Wooden spacers were inserted between each course of sleepers top to bottom to create a horizontal gap of about 25 mm.

As depth of manure grew inside the compound, liquid manure could escape between these slots and run down the outside of the wall. The liquid was caught in a concrete gutter, purpose-designed to lead the exuded liquid to an underground storage tank. From here, the liquid could be either pumped or tankered for field spreading.

Undoubtedly, some drying out of the manure mass occurred but opinions differ over just how much. In many such compounds the manure for 1 to 2 m in from the walls was clearly drier, but the effect did not reach throughout the mass. The lower slots in the sidewall often cease to run after a period and on inspection appear to have become blocked by a fibrous wad of material. Certainly some of these slot-walled compounds dry out fairly well and others much less so. The only study attempted to determine this question suggested that the amount of liquid collected after having escaped from the compound was very close to the winter's rainfall calculated to have fallen on the compound itself. If this is true, then the drying out of manure is probably largely due to evaporation from surface and sides of the compound during summer aided by the fact that rainfall, or its equivalent hydraulic load, has not been stored up in the manure. Quite a neat theory this but perhaps too tidy for reality.

(ii) Farm Storage Ltd of Otley, in Yorkshire, introduced onto the market in 1977 a variant on the slotted-wall principle. Their concrete-floored compound is surrounded by a wall 2 m high of concrete planks some 273 mm wide and 70 mm thick. These are positioned vertically with a 25 mm gap between adjacent planks so that a vertical gap is created.

An important feature of the design is that the floor is haunched with concrete all round the edge, creating a lake in effect but only some 150 mm deep. This gives the necessary layer of liquid upon which the mass of the contents can flow from the loading ramp to fill the whole of the compound evenly. Surplus liquid is expressed out through the sides via the vertical slots and collected in a gutter surrounding the compound on the outside at the foot of the wall. The liquid is directed to an underground collection tank, the size of which is determined by factors such as anticipated rainfall, size of dairy herd and intended frequency of emptying.

Opinions differ over sizing of this type of compound relative to the number of cows in the herd. Some maintain that 20 per cent less storage volume is needed, compared with conventional compounds, because of the free-draining nature of the design. Unfortunately, there is as yet no recorded data about this. Whilst undoubtedly drainage reduces the final total volume of manure in store by the end of summer, the practical problem will be as the store nears capacity, how much drainage will have taken place? This point of time will be at the end of the winter housing period, which will be before there has been any significant loss of water by evaporation. Hence, it is likely that a somewhat smaller reduction in capacity than the 20 per cent mentioned would be prudent.

(6) Straw Bale Compounds

This is a system developed at the MAFF Terrington Experimental Husbandry Farm near King's Lynn, in Norfolk, to provide pig slurry storage, eventually producing a material which can be handled like FYM. A rectangular compound (fig. 6) is made of walls comprising tightly packed bales three deep and four layers high. Each metre run of wall needs twelve bales. To prevent leakage of manure, joints between bales in the walls are covered over with old paper meal bags. The corner bales may be tied together with polypropylene string for stability. Loose straw and broken bales are shaken out in the compound to a depth of about 0·3 m and slurry is pumped in weekly from a tanker. More straw is added and gradually the compound fills up and because of drainage and evaporation a composting action is obtained.

A compound will last about a year depending on how wet the slurry is and after a further year, the compound can be shovelled

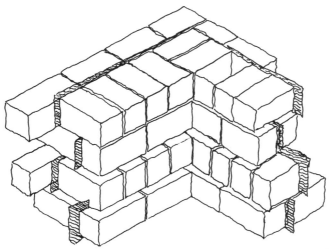

Fig. 6. Straw bale compound showing meal sacks between layers of bales.

up with the tractor loader and spread on the land. The compound walls are first of all burnt off to destroy the plastic baler twine which does not rot and would create problems with spreaders.

As a guide to sizing, a compound 30 m × 10 m × 4 bales high will need about 1,000 pick-up bales and will accommodate a year's slurry from 240 pig places, provided excessive dilution of the slurry has not occurred.

Like other compounds, this variety would probably require licensing of the site by the local or water authority under the proposals of the Control of Pollution Act 1974.

ADDITIONAL FACILITIES FOR USE WITH COMPOUNDS

(i) LOADING RAMP (fig. 7). The usual arrangement is for slurry and manure to be pushed up a ramp for loading into the compound. The ramp provides the necessary height to get over the retaining walls. Whilst a few ramps have been made of steel latticework with a steel, chequer-plate running surface, they demand fair skill and involve handling heavyweight metal. This sort of work is more in keeping with a good local agricultural engineer than the farm workshop. There are also problems in retaining adequate tractor wheelgrip when the chequer finish wears off after two or three years. Most ramps are built up with concrete block walls and a concrete runway. The most important features are:

- Good foundations. This feature is particularly important as the ramp must support a moving dynamic load of 2 tonnes or more (the tractor and scraper) *and*, since the ramp extends into the

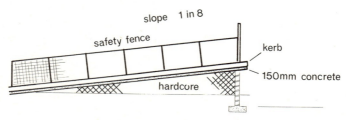

Fig. 7. Loading ramp for slurry compound.

compound, the foundations will have to contend with wet ground conditions.
- An incline not steeper than 1 in 8, preferably less. It is a common and embarrassing fault to build the ramp too steep, only to find the tractor cannot push a full scraper up the slope due to wheelspin. The necessary length of ramp can be surprising, *e.g.*, to gain 2 metres height requires a length of 16 metres. Thus a ramp is not a cheap nor particularly simple structure to construct and may well best be put out to a builder.
- It is particularly important to ensure that the transformation angle between the yard surface and the inclinded ramp surface is smooth and sufficiently gentle so that the scraper can follow accurately. Otherwise a wedge of slurry will be left behind each time if the scraper cannot follow the ground fully.
- The ramp surface should be level but with a tamped finish to provide wheelgrip. An alternative is to use wooden formers 50 mm by 25 mm during the laying of the surface concrete to form a herringbone of 'V' shaped pattern of slots in the surface finish spaced 300–450 mm apart. Although these fill up with manure and other waste, they do provide a grip even in icy weather.
- The sides of the ramp should have a coaming or kerb some 300 mm or more high to retain slurry shed out of the scraper sides.
- There should be safety rails along the sides and particularly at the top to prevent the tractor careering overboard into the slurry. It is too easy for a muddy gumbooted foot to slip off a clutch pedal when attempting to stop at the top of the ramp.
- Finally, the position selected for the ramp should be carefully considered. The aim is to allow speedy movements of the tractor scraper from buildings to, and up, the ramp. Although the most obvious location is end-on to the cattle buildings allowing direct tractor approach, very little extra time seems necessary where a 90° turn is necessary as several scraper loads can be deposited before the tractor need turn to sweep the accumulated slurry up the ramp.

(ii) STRAINER BOX (fig. 8). A principal aim of a compound is to allow the winter's manure to dry out sufficiently by late summer to facilitate unloading using high capacity solid manure equipment. The tendency for undisturbed slurry in store to separate into a bottom sludge, a middle liquor layer and a top crust has already been mentioned. If much of the middle layer can be

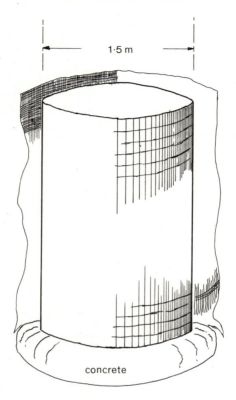

Fig. 8. Welded mesh slurry strainer box wrapped with finer mesh plastic.

removed, then the store contents will dry out better.

The strainer box or strainer chimney developed on south-west farms is a device to facilitate the removal of the liquid layer. At its simplest it is a cylinder of welded mesh about 1 to 1·5 m in diameter, long enough to reach from the bottom of the compound to 600 mm above the surface crust when the store is full. The chimney is secured to the bottom of the compound by attached concrete blocks and supported by four or more poles or pipes driven into the ground inside the chimney circle.

The welded mesh of 50 × 50 mm is then wrapped externally with a layer of Netlon or other plastic mesh having a hole size around 6 mm. This allows liquor to drain into the chimney or box, from which removal can be effected by either tanker or small electric pump for spreading to land.

The strainer box is best positioned at the deepest part of the compound which is adjacent to the ramp. Personnel access to the top of the strainer box is necessary to check liquid levels and hoses, etc, so a properly constructed catwalk complete with safety rails is usually required unless access from the ramp itself is feasible. Where the strainer is cheaply made, it needs protection from the pressure waves occurring in a compound which may distort or even push over the strainer. These waves are created when slurry is pushed off the ramp, falls as a lump into the semi-liquid contents of the compound, and builds up a local head or heap of slurry. This exerts a pressure to gradually push out the contents to the rest of the store. The presence of these wave fronts can be clearly seen as a pattern in the top crust radiating out from the loading point.

Lightweight strainers may be protected from being overturned by positioning three or four vertical posts, *e.g.*, sleepers, between the strainer and the loading point below the ramp.

The strainer system can be used to manage the height of the compound contents during winter. Usually the height is kept fairly high as this exerts an outwards hydraulic pressure and prevents in-leakage as well as providing a fairly free-moving store to encourage even depth throughout. Later in winter, sufficient freeboard is maintained to allow for any unexpected heavy rainfall, liquid being removed as opportunity for field spreading occurs.

An important practical point is that a depth of liquor some 500 to 600 mm deep should always be maintained in the chimney. This is to ensure that the top crust and bottom-most layer are always separated by liquor, as this allows liquid to move sideways throughout the store to reach the removal point in the strainer box. When warm weather returns in late spring, the maximum amount of liquor is then withdrawn to encourage drying out. A violent extraction rate is not the most effective; steady removal with intervals to allow percolation seems best.

The plant nutrient value of the liquor is pretty low and varies with the rate of dilution. A typical sample may contain 1–2 per cent dry matter and the equivalent of 0·8 kg N, 0·8 kg P_2O_5 and 1·8 kg K_2O per cubic metre.

Other variants on the same theme include a strainer chimney comprising interlocking concrete rings, the walls of which are perforated with 50 mm diameter holes fitted with a mesh filter.

Very robust, these are much more expensive at £200 each plus carriage, which may be costly.

Compounds have been satisfactorily de-watered using a three-sided box of sleepers against the bank. The box reaches to the bottom of the compound and the sleepers are spaced 25 to 50 mm apart to allow drainage from the manure. A single face or section of sleepers similarly spaced placed across the deepest corner of a compound has been quite successful too. All of these designs are a testimony to the inventiveness of farmers and their workers.

(iii) FENCING. The same sentiments apply to the compound or the lagoon—that both are very dangerous places for anyone and so should be very effectively fenced to allow access only by the holder of the key to the padlock on the gates. There really can be no short cuts over this if the traumatic and nightmare experience of someone's death in the compound is to be prevented with certainty. On a lesser moral but nonetheless cogent economic plane, a cow or horse lost in a compound can represent a noticeable loss of capital value—apart from the labour and fuss of an attempted rescue followed by the inevitable recriminations. So if only for peace of mind, top-class fencing is a must.

Very often the cost of fencing will more than exceed all other costs of constructing a compound, which fact often seems to surprise enquirers. The position, relative to the compound bank, of the fence merits some thought if it is not to interfere with unloading operations. The general tendency is to think too small in terms of layout which may bring a legacy of access headaches later in the life of the store.

Whilst mulling over the fencing problem it is useful to consider whether or not screening shrubs or trees can be incorporated in the grand design. This could do a lot for the local environmental quality and it is a point clearly considered and acted upon by continental farmers.

(7) Concrete Tank Structures

Concrete is an attractive material for use in manure and slurry work mainly because of its longevity, resistance to corrosion and its low to nil maintenance requirement. The concrete block is universally favourite for structures built by farm staff, and any tank will require reinforcement to contain the thrust developed when full to capacity. There is science involved in just how many steel reinforcing rods of what thickness should be positioned where in the walls. It cannot be too strongly emphasised that the services of a competent structural engineer should be used to specify reinforcement design.

Concrete block below-ground tanks

Two popular examples are reception pits for scraped slurry for transfer into a long-term store and as a storage tank for parlour washwater or 'gravy' from a compound.

A common misapprehension is that because the tank is below ground it will be supported by the sides of the excavation after backfilling, and so building practice need not be so exacting. In fact, from a structural point of view, it makes no difference whether the tank is above or below ground. The structure must be strong enough to bear all the loads upon it without cracking. Before the soil around a below-ground tank would have been compressed enough to begin supporting the tank walls, the tank would have cracked open long before.

From the slurry handling point of view, tanks are probably better deep and small in area than shallow and extensive but this conflicts possibly with considerations about ground-water tables. The deeper an excavation, the greater the pressure on the finished tank from ground-water trying to find a weak point through which to leak into the tank. Potentially, a concrete block tank, being made of many small pieces, is prone to leakage unless the highest standards of building practice are maintained and the internal face is rendered. Leakage into or out of an underground tank is bad news either for the farmer, who must then handle unnecessary volumes of ground water or for the environment which is subjected to unnecessary pollution of water supplies.

Underground tanks are often seen fitted with a reinforced concrete top with manholes for access. Wherever possible, this should be avoided in the interests of easy and complete access to the whole of the tank ('Murphy's law' being what it is—everything goes wrong sooner rather than later, especially where slurry is concerned). Blockages will need attention, and descending into a closed top tank is risky because of foul gas accumulation.

An open grid covered top is better. The grid can be in steel, concrete or wood and should be de-mountable to allow removal of pumps, etc or access for grab loaders for cleaning-out purposes. Obviously, the grid should be designed to safely carry the loads it must carry. It is important to decide whether or not tractors and trailers and/or lorries may need to pass over the grid. For lorries, the EEC maximum axle weight is just over 11 tonnes but this is deceptive. A lorry which arrives fully laden in the farmyard may discharge the rear half of its load and then manoeuvre over the grid either to adjust its position for further unloading or on the way out of the farm. In this case, axle weight may reach 16 tonnes.

Underground tanks of rendered concrete block construction are used in most livestock housing incorporating slatted floors—it is very rare to find poured concrete used for this job. Contrary to popular opinion, the floors of all such slurry tanks should be laid level and *not* with a fall. This is so that urine and faeces do not separate out, the urine running to the low end leaving the solids behind to dry out, adhere to the floor and so begin to form a problem. A fundamental rule is that liquid should remain mixed in with the slurry mass held in the tank by the sluice gate. This presumes, of course, a leakproof tank. Given a level floor, the main concern is to ensure that the sluice gate fitted at the discharge point is watertight and so does not allow liquids to drain out. This is a matter of good design of the sluice (it should not bend under stress), careful detail workmanship in building and regular inspection and maintenance during use.

Empty under-slatted-floor tanks should normally be filled with enough water to cover the floor 50 to 100 mm deep. This prevents the first solid manure from adhering to the floor. This addition of water initially may not be necessary in a fattening-pig house, especially where pipeline feeding is practised. Extra water to 150–250 mm deep is vital under beefstock and may be required under dairy cattle. The decision about how much water is best determined by experience of the particular livestock building. The dry matter of the muck produced is a function of the type of rations being fed. Thus management must be exercised of the slurry system just as much as the rest of the livestock enterprise. Changes in feeding regime, *e.g.*, a reduction in the amount of silage fed upon the inclusion of hay in the ration, will demand management decisions about the slurry system.

It is probably simpler here to comment separately on the various livestock systems.

(i) PIGS. The slurry tank or channel is best if width is 1–1·5 m. Greater widths run the risk of 'meandering' flow of slurry and the gradual build-up of islands of solids leading eventually to blocking. Depth of channel may reach 1 m, being determined by number of pigs housed and the length of storage period required as well as site conditions (high water table or shallow soil over rock, for instance). Rather than go for deep channels to give lengthy storage, it is much better to size channels to give 2–3 weeks' storage and provide the long-term storage away from the buildings in a lagoon or tank.

If channels are very long, it is preferable to divide up the length in sections of 20 m or so by a sluice gate followed by a step down of 0·5 m in the floor of the channel (fig. 9). This encourages rapid discharge and some re-mixing of the slurry as it travels to the final discharge sump, tank or removal system. This final sump should be large enough to accept all the volume from one section at least, to avoid backing up along the channel during discharge.

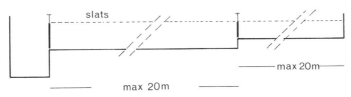

Fig. 9. Very long slurry channels should be divided by sluice gates every 20 m.

Removal can finally be by pump or vacuum tanker from the sump or onwards by pipe to a lagoon. Such underground pipes should be laid in straight lengths with inspection/rodding chambers every 35 m and at bends. Pipe diameter should not be less than 150 mm and laid to use the maximum fall available on the site, with 1 in 50 a minimum.

(ii) CATTLE. Relatively few dairy units are to be found on slatted floor accommodation. Where installed, the slats usually extend to the cubicle passageways and the feeding area. It is here that spillage of forage may cause problems, but this can be alleviated by diversion of the parlour washwater to this area of the under-slat storage. The occasional 'dropping in' of 5 m^3 of water from a tanker can assist freeing of areas of more solid material, but it is very important to keep on top of the situation by regular inspection as established slurry solids are very difficult to dislodge without lifting slats. Good design will provide a number of access points around the building where, via sumps, slurry can be removed or water added.

There have been developed livestock housing systems for intensive production of beef where all livestock accommodation is on a slatted floor (fig. 10). The best known system was evolved in Ireland, yet under UK conditions the slurry handling has not always been successful. The inherent problem with beef is the rather higher dry matter of the manure, especially where straw forms part of the diet.

The original systems used silage, often low in dry matter, as the bulk diet. Moreover, the climate is rather wetter in Ireland. These two facts are probably basic to the success of the system and were somewhat overlooked in the UK. The depth of slurry cellar is usually considerable—2 m or more to give extended storage—and the system relies upon removal by pump or tanker of slurry via access sumps or pipes around the building. Despite running in water to a depth of 300 mm before stocking up, supplemented by diverted roofwater during the season, a number of UK farmers

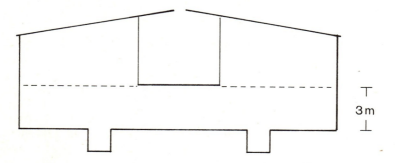

Fig. 10. Beef building on slats—depth of cellar and emptying sump depends on site conditions and length of storage period needed for slurry.

have had great difficulty in clearing out cellars. Ideas tried mainly unsuccessfully include:
- Tanker loads of water 'dropped' into cellars.
- Pump tanker used to suck up and jet back slurry into the cellar compartment by compartment.
- Pumping slurry at maximum rate from the unloading sump to the far end of the cellar.

- A hired industrial air compressor to supply air to a perforated tube positioned along the length of the cellar floor.

What seems to be a telling mistake is to unload a little during the fattening period, as inevitably it is the liquid fraction which comes easiest leaving a disproportionate amount of solids behind. Because the building is full of stock, the operator is probably unaware of the longer term problem he is creating.

It is perhaps not altogether unconnected with the possibility of unloading difficulties that an Irish manufacturing company marketed a tractor-pto-driven mechanical mixer designed to reach between slats into the slurry below, comprising a right-angled drive shaft carrying a cross- or star-shaped paddle of four arms which can fold so as to pass between the slats. Sometimes one slat must be removed to allow enough room.

A more practical solution is to modify the building design to allow tractor access below the slats and to encourage maximum drainage of liquid. This can be achieved by spaced strainer hatches at the tractor access points assisted by strainer chimneys or boxes at intervals under the slats. This would allow escape of liquid to a collecting tank for periodic disposal. Manure could then be removed as a solid or semi-solid using the tractor front bucket.

(iii) POULTRY. Older designs of battery houses incorporated belts to catch droppings. At regular intervals, sometimes automatically, the belts were scraped clean as they passed over a reception pit at the end of the building. Removal from this pit may be either by mechanical conveyor, or by tanker or auger, after water has been added. Battery hen muck comprising feathers, waste food and spilled water will vary around the 20–25 per cent dry matter level.

Of recent times, battery cages have been mounted in staggered tiers (fig. 11) over a concrete floor. Known as the deep pit system, droppings, etc fall direct to the floor where they gradually dry out due to the passage of exhaust ventilating air down over the hens, over the manure and out through slots in the side walls supporting the building and containing the droppings. Emptying is by means of tractor bucket, the manure being fairly dry (75 per cent dry matter) after a period of a year or longer in the deep pit.

The egg-laying poultry industry has increasingly turned to this system because of its convenience and because of the reduced smell level compared with wet poultry slurry. The most important management detail is to keep water out of the pit. This implies a

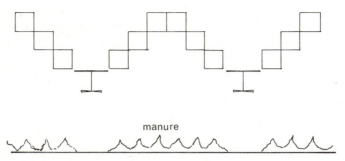

Fig. 11. Battery cages and deep pit poultry manure storage showing optional air recirculating fans.

soundly constructed floor and, above all, daily checks to see that the drinking water equipment does not leak. Serious water leaks can result in a (molten) mass of slurry gradually building up against the tractor access doors and eventually bursting them as the manure escapes downhill.

Tanks constructed of concrete sprayed in situ

This interesting form of construction has been marketed since 1979. A circular tank can be formed by opening out a roll of special welded steel mesh to form a circle of the required diameter up to 2 m high. Onto this is secured a made-to-measure one-piece liner of butyl rubber 1·5 mm thick. This structure is then reinforced by the application of concrete using portable 'wet guniting' equipment.

About 75 mm thickness of concrete is built up, which cures to allow filling of the tank seven days later. Thus a very quick installation service is possible. One of the reasons for the speed of erection is that no concrete base is required. The site has merely to be levelled and then blinded with a 25 mm layer of sand to protect the butyl liner from sharp stones. Sizes vary from 227 m^3 to 727 m^3 with diameters from 6 m to 11 m. Costs vary with the number of fittings, such as drain valves and inlets specified, but a range around £15/m^3 to £27/m^3 for the smaller sizes can be expected.

This type of relatively shallow large diameter storage tank is probably more suited to very thin or dilute slurries than to whole slurry. Separated liquor would be suitable too. The reason for caution over storage of whole slurry is that, should sludge deposit on the bottom of the tank, only hydraulic means of removal could be used lest mechanical agitation damaged the liner. Removing sludge from a small capacity and diameter tank is less of a problem.

In practice, it is a sound precaution to homogenise slurry thoroughly before adding to the tank and to recirculate the contents regularly, particularly during unloading. In this way sludge should not be a problem.

Concrete panel stores

These above-ground structures are made of concrete panels, with edges designed to lock with adjacent panels. The round structure is supported by external bands of steel rod tightened to lock the construction into a rigid entity (Plate 2). The floor slab needs to be laid level accurately so that panels locate well; the inside wall surface is usually painted with bitumen or other flexible material to both protect the concrete from attack and to make it leakproof.

The main maintenance requirement is to ensure the exterior steel bands do not corrode over the years. Failure of only one of these bands could lead to consequential failure of others and so a major store collapse.

Whilst this type of store construction is cheaper by some 15 to 20 per cent than others, good looks are not a strong point—some would even class them as an eyesore.

Stores made of concrete pipes

A novel form of store construction is to use numbers of 600 mm diameter concrete pipes which are stood on end and bedded into a strip foundation of concrete. By positioning the pipes as close together as possible and sealing the vertical joints, a wall is created which will retain slurry. A floor of 100 mm concrete is laid within the ring of pipes. The depth of the store is limited by the vertical length of the pipes, usually 2 m. Pipe costs are high unless rejects are purchased and transport costs will be high because of their weight and bulk.

Perhaps the unique feature of the system is that the store shape can be round or almost any rectangular layout. Space within each pipe can be used for storage, although the mind boggles at the fiddle to extract the slurry from within the pipes eventually. All in all, an unusual method of store construction but not one to give sleepless nights to the sellers of other types of store in the UK.

(8) Above-ground Steel Slurry Stores

These circular metal stores have become a familiar sight on many farms and are probably the most popular form of purchased storage available. The attractions are longevity, low maintenance requirement, a convenient handling systems when well-managed and, above all, the fact that these stores offer complete containment with no chance of leakage. This last attribute can weigh heavily with the farmer who has had a brush with the law over water pollution and who cannot afford a repeat problem. Furthermore, most manufacturers directly or through their local dealer offer a complete installation service from initial design survey, through excavation and construction to the handing over of a complete store fitted with the necessary machinery to provide a fully working system ready to go. For the busy farmer this, too, can be a decided advantage.

(i) SITING. Apart from deciding the actual size of the store, the greatest care is needed over siting. Because these are expensive structures its obvious that the site needs to be a permanent one and so this should be chosen from two stand points: the internal farm requirements and the local environmental standards. Usually, the position of the store within the farm buildings complex is dictated by existing space and access to it from the livestock housing. Whenever there is the temptation to site the store across the gable end of a building, it is time to pause and ask what likelihood there is of further expansion of the particular set of livestock housing. To answer this, the timescale should be looking forward at least a decade, not just to next year. This short exercise may just prevent a large and costly store being sited in the wrong place, a mistake which is more common than might be expected.

Having identified a possible site within the farm steading for the store, it is well worth the trouble to walk around neighbouring vantage points from which the farm may been seen to try to picture how the scene will look after the installation of what may be a large tank. It is just possible that a slight change of siting may be seen to be of long-term benefit to the local environment or that the planting of some trees may help the situation. Such a minor precaution need take little time, needs to be done but once and yet can do so much for the countryside panorama. The green or blue colouration given to the various makes of tank can be a striking addition to the variety of colours contributing to the farming scene and certainly at least some tanks have been sold due to a strong preference for one colour or the other.

(ii) CONSTRUCTION. The individual plates of steel are bent to the relevant curvature and then receive a coating of coloured vitreous enamel on both sides, so preserving the steel almost indefinitely. Sheets are bolted together at their edges, thicker sheets being used in the lower rings of the structure to withstand the greater pressure from a filled store. The joints between sheets are coated with a mastic sealing compound to ensure a good seal and to accommodate any relative movement between sheets due to temperature effects—the sunny side of a tank gets much warmer than the Northern face. A useful tip during construction is to ensure the erection gang are using sufficient mastic.

The ring of steel walling is carried on a specially designed base foundation which incorporates a locating ring to accept the bottom of the wall and to provide a good liquid seal. The foundation and floor of the store must be laid accurately if the store is to assemble without problems and careless builder's work can result in the whole floor having to be relaid.

The floor usually incorporates an unloading pipe connected to the adjacent reception pit so that slurry can be recirculated from the store, mixed and returned to the store or to the spreaders during unloading.

The reception pit may be of concrete block construction but most often is a miniature version of the main store sunk into the ground. The vitreous finish is proof against corrosion and leakage—two considerable advantages especially in view of the ease of construction. Two sizes of pit are generally offered with diameters in the range of 2·5 m to 3·0 m and depths around 3 m. Table 15 gives an example of the range of sizes of stores available from one of the several manufacturers of this type of store.

(iii) ACCESSORIES. A number of items of equipment are necessary to facilitate the operation of a slurry store, and although the pump is the heart of the system it will be dealt with elsewhere (page 119) amongst other machinery. Hardly an accessory, more of an essential and deserving a mention, is the access ladder to enable inspection of the tank contents. In earlier years, some farmers merely leaned a ladder against the side of the tank to climb up: this is a dangerous practice to be avoided. Manufacturers supply a purpose-built ladder cum inspection platform complete

TABLE 15. Capacity of Circular Slurry Stores, (Cubic Metres)

Height in metres	Diameter in metres						
	9·1	10·7	12·2	13·7	15·2	16·8	18·3
3·6	235	321	418	531	656	785	920
4·8	313	427	557	707	873	1,041	1,227
4·6	391	533	696	883	1,090	1,297	1,534
6·8	469	639	835	1,059	1,207	1,553	1,841

with all necessary safety rails. This allows comfortable and safe access to the top of the store and more than one such platform may be needed.

Most interest lies in those accessories which are available to keep the contents of the store well mixed. They may be divided between those using hydraulic, pneumatic or mechanical principles of mixing:

Hydraulics. Slurry from the main store is run back into the reception pit from whence it is picked up by the chopper mixer pump and returned at maximum flow rates (80 l/s) back into the main store. With the larger stores, simply recirculating at this rate will not overcome any crust formed on the surface and so jetting kits are supplied. These comprise pipework to reach up to three positions around the store circumference and terminate in a hydrant jet or nozzle. This can be directed at will to blast any crusted areas with a high speed stream of slurry which will also entrain some air into the tank contents. Such jetting kits can be effective but may need a considerable time to overcome serious crusting of the surface. To supply such large volumes of pressurised slurry, the pump demands a fair power input and may provide a very full load for a medium-power tractor on the pto shaft. About 25 kW may be required, and so prolonged jetting does have a cost for fuel and tractor hours apart from the supervision labour costs.

A variant on the hydraulic stirring theme is to lead all the flow from the pump to a group of pipes permanently installed at the side of the store. Typically, three pipes may be used each fitted with a nozzle variable in direction controlled by a rod reaching to the top of the store. Each of the three pipes has its nozzles at a different depth with the aim of mixing more thoroughly than if slurry is only returned onto the top surface. Again, high volume and pressure are necessary to obtain the best effect.

Pneumatic. There have been a number of designs of agitation equipment for large slurry stores based on compressed air. The main differences lie in the design of the outlet used and submerged at the bottom of the store. Fairly large air bubbles are necessary because fine small bubbles tend to act as a lift for fibrous material which then tends to collect in the top crust, so making the problem worse.

Individually designed bubble generators have been tried made of plastic. About the size of a five-gallon drum and incorporating a venturi shape, the idea is to generate a large bubble which will sweep upwards carrying a blob of slurry towards the surface. Just as successful in operation has been a recently introduced design thought up by a farmer. The air supply pipe is secured to the floor of the store and terminates in an outlet covered with a very simple diaphragm one-way valve so that air can escape but slurry cannot flow back into the pipe with the risk of blockage.

There are three such bubblers positioned around the store and so that only a moderate sized compressor is needed to economise on motor size, all the air output is switched to one bubbler at a time, each in turn. This is achieved by a sequencing air diverter valve supplied with compressed air from a 3 kW electrically-driven compressor. The whole unit is controlled by a time clock so that the farmer can select the length and number of periods of operation each 24 hours. This automatic control is practically very important in that the store agitation can continue without relying on the farmer's memory: on a busy farm this is invaluable.

The main lesson from pneumatic agitation is that air bubbles can keep a store mixed provided the system is used from the beginning of filling. Air bubbles have been seen to quite fail to tackle a really well-established crust on cow manure, hence the advice to use the bubble system at frequent regular intervals.

Mechanical. A certain amount of mixing of a store can be achieved by a floating agitator or aerator, but these will be discussed later under aeration devices.

Mechanical mixing in above-ground slurry stores has been achieved by electrically- or tractor-driven propeller mixers. In principle they are a long shaft to transmit power to a propeller. Tractor-driven models typically have four blades of 500 to 600 mm diameter, but electric models are usually smaller to avoid unduly high motor power requirements.

In Europe there are designs which are mounted on the tractor rear 3-point linkage and can be lifted up and over the edge of the store and by an ingenious drive path, the tractor pto shaft powers the propeller inside the tank. There are on the UK market a few designs of propeller agitator which are positioned near the bottom of the store and driven via a shaft passing through the wall which is reinforced at this point. The propeller can be varied in direction to aid mixing. A very high standard of engineering construction and tolerance is probably necessary if vibration is not to be transferred to the main tank structure with possible defects occurring.

Management of above-ground stores

The guiding light should be to prevent problems arising rather than spend much more effort trying to rectify them afterwards. The following points are worth bearing in mind:

- Where possible, opt for a taller narrower store than one shallower but wider.
- The slurry must not be too thick. Pumps will handle manure at quite high dry matters, over 10 per cent, but the problem is that the manure will not run to the pump intake, so some dilution is necessary. The amount of added water should be minimised as it is occupying costly storage space. In practice, adding the dairy washwater will usually dilute cow slurry enough and pig slurry is normally wet enough at typically 6 to 7 per cent dry matter.
- The amount of bedding and long fibrous material should be strictly controlled and as much as possible kept out of the slurry system.
- A lookout should be kept for foreign objects at all times.
- Sand should not be used for cubicle bedding as it must eventually be dug out of the store by hand.
- When a fresh day's scraping of slurry reaches the reception pit, the contents should be thoroughly homogenised by recirculating for 30 minutes through the chopper mixer pump before transferring into the store.
- The store contents should be recirculated weekly. The time devoted to this will depend on size of store, how full it is and the presence or absence of surface crust. Two hours is a minimum. If an automatic system is available it should be set up in the light of store conditions lengthening operating periods as the store reaches capacity.
- Stores should be *completely* emptied out at least annually.
- No reception pit should be entered unless one is wearing a lifeline manned by two assistants on the surface lest foul gases have

collected. No personnel should work unaccompanied at the top of a slurry store. Where staff must work above a depth of slurry, life-lines and life-jackets should be worn and careful supervision exercised. There are few second chances when floundering in slurry.

Provided these few golden rules are religiously followed, a convenient slurry handling system will be enjoyed. Transgress and the dividends can be frustrating, annoying and expensive too. Stores have been seen where a crust formed unheeded and the store could not be completely emptied. In one case the crust fell and rose on the second and third years' slurry now adorned with bushes and even saplings. The cure was to empty out liquids, unbolt the sides and procure the services of a digger loader inside the tank. This was not only time-consuming and expensive but, worst of all, the unfortunate farmer confided, was the incessant leg-pull from friends and neighbours.

Chapter 6

SLURRY AND MANURE HANDLING

THE METHODS, machines and systems by which slurry can be handled are legion and some items are difficult to classify. However, a very simple split is between the handling of slurry in and around buildings (usually, but not always, by fixed plant) and the handling of slurry to the field.

Clearly the consistency or dry matter of the slurry will primarily determine which types of machinery are suitable for use in a particular situation. Fig. 12 is an attempt to illustrate this and so avoid a necessarily lengthy description. It goes without saying that the various descriptions of the slurry gradually change as the dry-matter percentage increases and not that, for instance, 'thin' slurry becomes 'thick' at exactly 12 per cent dry matter. Similarly, the scope indicated along the bottom axis of the slurry-graph for various machines should be regarded as an indication only—the presence of long material in the slurry can greatly influence these indications.

Study of the slurry graph in fig. 12 plainly shows that on occasions water may have to be added to slurry to reduce the dry matter sufficiently for a particular machine or pump to be able to handle the material. Excess dilution costs money for the water and for the extra work involved in disposal. So the obvious question is how much water need be added to change initial dry matter of the slurry to a given final level. This can be worked out mathematically but a more convenient 'calculator' is shown in fig. 13, where by placing a straight edge across any two columns, the corresponding figure in the third column may be quickly read off. The example illustrated shows that manure initially at 10 per cent dry matter would require the addition of around 570 litres of water per cubic metre of manure to reduce dry matter to 6 per cent—a typical level for feeding into an anaerobic digester, for example.

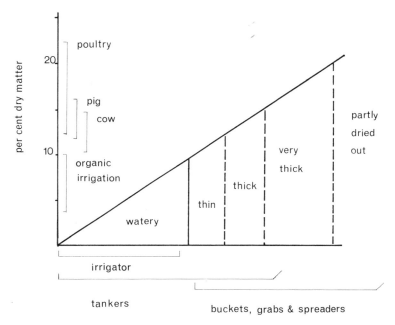

Fig. 12. Slurry graph showing the relationship between dry matter, consistency and suitable handling equipment.

HANDLING IN AND AROUND BUILDINGS

(i) Mechanically Unassisted Handling

This section refers only to slurry which, being a semi-liquid, can be inveigled to flow unassisted altogether. An alternative ploy is to use hydraulic means or flushing to move slurry.

Overflow slurry channels

As mentioned earlier, slurry channels should be laid with a level floor. It has been found that by installing a small lip or dam at the end of a channel, a continuous discharge of manure is obtained. Fig. 14 shows the general arrangement. The 150 mm high lip retains a liquid layer on the base of the channel, which is filled to the lip with water initially before livestock occupy the pens served. This prevents solid muck from adhering to the floor and perhaps drying out before there is sufficient liquid present. Having established the bottom liquid or lubricating layer, subsequent additions

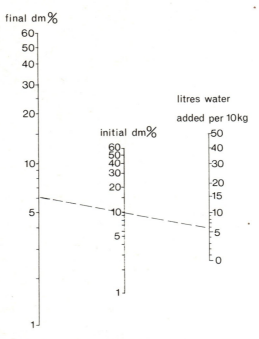

Fig. 13. Nomograph showing the amount of water needed to change the dry matter per cent of 10 kg of manure.

of slurry which would ordinarily pile up can only do so to the extent that the slurry develops sufficient static head or thrust to slide along the liquid layer and fall over the lip. In practice, it has been found that the top surface of the slurry assumes a gradient of around 1·5 per cent usually and occasionally 3 per cent.

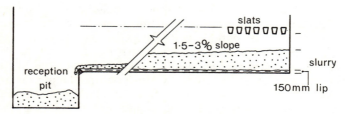

Fig. 14. Overflow slurry channel.

SLURRY AND MANURE HANDLING

Extra water may be added if the building has been without stock for a time to make good any evaporation losses. Since the system depends upon the retention of liquid in the bottom of the channel, it is obvious that quality construction is necessary to avoid any leaks.

The advantages of the system are:
- No supervision required.
- Maintenance requirement almost nil—a thorough clean out once every three years or so.
- Slurry discharges continuously without need for agitation.
- Reduced building costs as only shallow channels are required.

NB—the system is *not* recommended for animals producing high dry-matter manure (greater than 12 per cent), of which the best example are hay-fed cattle or where trough design allows a lot of food wastage onto the slats.

Channel sizes

For cattle, the width is normally between 1·5 and 4 m (for pigs 1 to 1·5) as greater widths may lead to meandering flow. Maximum length of channel seems to be about 30 m.

The depth of channel is determined by the length and the inclined surface of the slurry which for design purposes ought to be taken

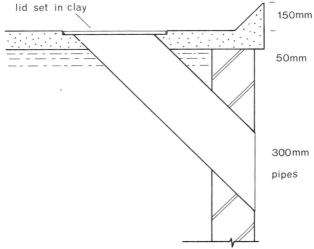

Fig. 15. Overflow slurry channel—detail of cleaning out drain.

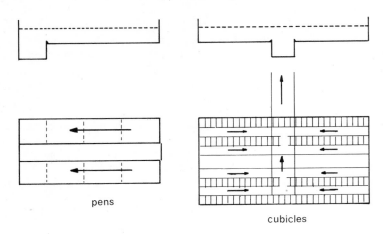

Fig. 16. Overflow slurry channels—alternative arrangements.

as 3 per cent. Thus the distance from the underface of the slats to the finished level floor of the channel will be the sum of:

150 mm—depth of lip.
250 mm—freeboard between top of slurry surface at its highest point (opposite end from lip) to the slats.
3% × length in metres.
For example a channel 15 m long, depth will be

$$150 + 250 + \frac{3}{100} \times 15 \times 1{,}000 \text{ mm} = 850 \text{ mm}.$$

To aid the infrequent cleaning out of the channels, a drain may be incorporated in the end to by-pass the lip as shown in Fig. 15.

The system

Overflow channels may be used under animals to collect and deliver manure either direct to a storage tank or to a sump for removal by tanker or pump. The primary channel can, of course, deliver into a second overflow channel, to remove manure from the building to an outside storage facility. The depth of such secondary channels is worked out as shown above for primary overflow channels. Some of the alternative arrangements are illustrated diagrammatically in fig. 16.

The information on overflow slurry channels has been put together, after considerable experience, by Mr Jan Čermàk now of ADAS but formerly with the Scottish Farm Buildings Investigation Unit at Craibstone in Aberdeen, which has produced a useful leaflet on the technique. Strangely, the principle has been independently 'discovered' by more than one farmer without knowledge of the Scottish work.

One Yorkshire pig farmer made his discovery accidentally. The slurry pump broke down for several days and the sluices were raised on more and more channels as time passed so that maximum storage volume under the pigs and in the transport channels leading to the large pump sump could be used. After the pump had been repaired and the system was back to normal, all sluices were closed again to return to a fortnightly discharge cycle. One sluice was inadvertently left out, which was not discovered until six months later. The particular channel had never looked to need emptying during all this time and as several workers were involved, each thought a workmate had recently emptied the channel. It was then realised that the channel was self-emptying and the bottom of the old sluice gate guide was acting as a bottom 'lip'. All sluices were removed and the system operated as an overflow type every since.

'V'-shaped slurry channels

Work done at the Department of Environment's Building Research Station, Watford, early in the seventies looked at the flow properties of slurry. A special piggery was built in which the slurry flow system could be varied in angle of inclination. The upshot of this work was a design for a V-shaped channel having an included angle in the bottom of $45°–60°$. The channel proved effective when used in conventional mode with a sluice gate to store slurry. More interestingly, the channel proved able to discharge slurry even if left as an open channel laid with a zero fall. The sequence being that urine and faeces built up and blocked the channel until a sufficient hydraulic head of slurry built up behind the blockage to burst through. Hence the 'continuous' discharge referred to was seen as an irregular flow from the end of the channel. The channels may also be used as a flushing system whereby liquid pumped around the system will sweep out manure effectively. With a time control on the pump, manure can be automatically removed several times a day, with benefits to the environment within the pig house. The design was patented by the NRDC and the channels are produced in glass reinforced cement

(GRC), which is a modern material having the longevity of concrete with strength *plus* a degree of flexibility so that quite thin sections can be used compared with the same structure in ordinary concrete.

Although the experimental work was conducted with a channel 15 m long, all the indications were that the flow of slurry would be maintained at greater lengths. The researchers commented that flushing of the channels was very effective as a means of cleaning and only low flow rates were needed: 700–900 l in total at 45 l/s.

At present very few piggeries have been equipped with this type of channel, probably because of the high cost (around £25/metre). A 500-place fattening piggery completed in 1979 by the Ministry of Agriculture at their Terrington Experimental Husbandry Farm near King's Lynn, in Norfolk, incorporates 'V' channels, the performance of which is to be monitored for a lengthy period of time.

Hydraulic flushing

At first glance, using water to flush away manure seems an attractive, almost natural way of removal. Such is probably the sub-conscious pre-conditioning we all have received after a lifetime of pulling the chain. The Romans certainly used a flush removal lavatory in which the user sat above a running stream which was ducted underground below the houses of the influential and ultimately discharged outside the walls of the city. Some three years ago, excavations in the city of York discovered such an underground stream and the archaeologists found much of interest in the way of artefacts trapped in the sediment of the water course.

This vividly illustrates three needs of a good flushing system: a plentiful water supply; good design to ensure flushed material is transported efficiently and not deposited, by maintaining high enough liquid velocity; and last but certainly not least these days, a satisfactory disposal system. The first two requirements are not insuperable, but the last is nowadays very problematical. Large volumes of foul water are not easy to dispose of onto land in winter without fear of pollution. Hence a large storage capacity is needed. In the interest of economy, recycling of water is necessary and consequently the reception lagoon must be designed so that sedimentation can take place and cleaner liquid piped to water storage lagoons.

The reception lagoon silts up in time and so two such lagoons are required in parallel, the first to sludge up being allowed to dry out before cleaning out whilst the second is in use and vice-versa. The water storage lagoons, unless very large in area, can give problems of odour in summer or when the liquid is irrigated to land, and the smell problem can become an albatross around the neck of a flushing system. The problem is of anaerobic conditions giving rise to the formation of odorous compounds which are released during irrigation or even when liquid is recycled for washing. The practical cure in UK conditions, where odour is a problem, is to resort to mechancial floating aerators which are not cheap to instal (£2,500 each depending upon costs of the electricity supply wiring; 1979 price) and moreover represent a continuous running cost.

From the design point of view, effective flushing seems to need the establishment initially of a miniature wall of water say 75–100 mm high. To retain its kinetic energy as long as possible, it is helpful if the concrete area to be flushed is laid at a steady fall. For practical reasons, a reasonable incline would be something between 1 in 30 and 1 in 40. Eventually the flushed area must terminate in a collection channel which should be wide enough to prevent overshoot by flushing water which will approach at right angles. A width of 500–600 mm should suffice and depth 200 mm or more.

These guidelines resulted in a successful system for a large cow herd where, for example, a passageway between cubicles for 66 cows measured 2·4 m wide and 37 m long. 3·5 m^3 of water discharged from a tank into the passage satisfactorily swept out slurry—without bothering the cows at all. To do this took the cowman just over a minute morning and evening taken to operate the tank valve for twice a day flushing. Where a tip-tank is used operation can be automatic entirely. This very low labour requirement is the main attraction of flush cleaning but the system is used in America much more than in UK, probably because disposal via organic irrigation rain guns fits in with the great need, in some areas, for irrigation to guarantee grass production.

It is likely that in only very rare circumstances would the system be worth considering under UK conditions, having regard to the tightening of environmental conditions and regulations which seems inevitable in the future.

Comments so far have applied to longitudinal flushing of passageways, but some work has been done at NIAE on the flush washing of cubicle passageways across their width. The combination of water volume and pressure required has been investigated, but the work is not yet at a stage of rivalling longitudinal flushing.

(ii) Mechanically Assisted Handling

By far the most popular device is clearly that maid of all work—the agricultural tractor—on to which may be hung a variety of attachments to tackle so many different jobs. It is not the intention here to launch into a treatise on agricultural handling machinery, as it is assumed that readers are well aware of the variety of tackle available. However, a few random comments are offered on some items.

Handling solid materials
- Tractor-attached front muck forks have become larger over the years as the capacity of hydraulic systems has grown too. Tear-out forces of over 1·5 tonnes can commonly be achieved and a one tonne or greater payload on the fork is possible, especially where a top grab cage is fitted to the loader. Power steering on the tractor is highly desirable if the driver is to keep up a good work rate throughout the day. Where high work rates are achieved, the tractor clutch and transmission will have quite a hard time, especially where the driver is mechanically unsympathetic. It is not unusual for a new clutch to be needed after each winter's clearing out of livestock yards, etc. Some idea of work rates and load weights can be gained from Table 16, although when working in buildings, output is largely determined by the amount of room available for manouevring.
- Much less wear and tear on the tractor occurs when muck loading is carried out using a tractor-attached slew loader. More expensive models can slew through 360°, which is sometimes an advantage, and although fork loads tend to be smaller than for tractor loaders, their overall work rate is about the same because of their higher speed of digging. They are, however, sometimes limited by the reduced headroom in older buildings. It would be fairly unusual for a slew loader to be bought solely on the

TABLE 16. Handling Farmyard Manure

Machine	Av/wt of forkful (kg)	output (t/h)
Tractor front-end loader	355	36
ditto heavy duty + top grab	477	43
Slew loader	240	37
Crawler loader	767	62

PLATE 9
A four-wheel drive rough terrain fork lift truck at work loading FYM. The large capacity manure fork is fitted with a top-acting grapple cage.

Sanderson (Forklifts) Ltd

PLATE 10
Cleaning out litter from a broiler shed. This 4-wheel skid steer shovel is ideal having a large payload bucket and able to penetrate into the many awkward corners. Because of the low headroom, the top of the substantial driver protection cage has necessarily been removed whilst in the shed.

Clark International Marketing, Bobcat Div.

PLATE 11
A little motorised assistance with mucking out can greatly reduce the hard work.
British Farmer & Stockbreeder

PLATE 12
Automatic mechanical scraping makes a clean job.
Alfa-Laval Ltd

PLATE 13
Not a puzzle contest but a tractor-pto-operated slurry pump before lowering into the reception pit. The auger cutter and recirculating pipe can be seen at the left (bottom) end.

PLATE 14
A close-up view of the auger cutter of a slurry pump.

basis of muck handling, but their all-round versatility for so many jobs makes them popular.
- The big development in the last five years (1975–80) has been the take-up in agriculture of fork lift trucks for materials handling. Known as rough terrain fork lift trucks (RTFLT) (Plate 9), a variety of sizes offering two- or four-wheel drive are now available with a price range from £7,000 to £25,000 (1979 prices) depending upon power, drive arrangements and number of attachments. Buckets, forks and grabs are available with large capacity, two tonnes or more. The load weight is often limited by the density of the material. Fitted with power steering, change-gear-on-the-move transmissions, instant reverse and with good driver visibility, high work rates can be achieved.

  The skill in driving is not so much to avoid collision with building structures, etc but to deposit the solid manure into the spreader accurately and correctly, *i.e.* to tease out large lumps so that they do not overload the spreader mechanism when unloading in the field. More care is needed with the type of truck having a vertically telescoping mast lest the mast top fouls overhead obstructions. This is more of a problem with the two-section masts than the three-part versions. Obviously, greatest traction is available from true four-wheel-drive models, and although it is surprising just what difficult ground conditions an ordinary two-wheel-drive RTFLT can cope with, they can get bogged down and then may be something of a challenge to recover. The blame lies not so much with the truck as with the driver who, because of the truck's inherently good traction for two-wheel drive, may be led astray and attempt ground conditions too difficult.
- For very narrow passageways there are available small four-wheel-drive skid-steer loader trucks (Plate 10). Steering is by skidding round with one set of wheels locked, neither of which is steerable. This sounds crude but is nonetheless surprisingly effective and tyre wear is much less than might at first be expected. Noisy to drive, they are very quick in congested spaces and being narrow can be used to scrape out passageways. Slightly larger models have become popular with contractors clearing out broiler sheds. The easy access areas are cleared out with a tractor and large foreloader, leaving the corners to the skid-steer shovel. Altogether there are twenty-seven models on the UK market, costing from £4,500 to £23,000 (1979) with lift capacities between 230 kg and 1,680 kg.

- For narrow passageway clearing as in some types of piggery, two-wheel garden tractors fitted with a bulldozer blade can be used. They are certainly quicker than a shovel and wheelbarrow but are restricted to pushing and have no loading capability. The practical points to watch for are ease of starting the engine, easy controls and correct balance when fitted with a blade. A combination of unbalance during use and awkward controls can so easily result in the tractor taking charge whilst working and attacking the building, with damage to both!
- For scraping large areas, there are a host of tractor scrapers available. In the wrong hands, these devices can hand out a deal of punishment to buildings and fittings as well as effectively scraping up manure and slurry. Models are available which can be used backwards or forwards, as they automatically roll over when direction of travel is reversed, which is sometimes convenient. A good freeboard to retain slurry and spring-loaded sideboards are useful. Good heavyweight construction is an advantage with a design allowing simple replacement of the scraping edge, usually rubber. Finally, it is worth checking that the scraper scrapes cleanly—some don't.
- The disadvantages of tractor scraping are that it can only be done whilst the cows are out of the accommodation, which is a nuisance. The job cannot be done too quickly or else slurry is flung up and about by the tractor rear wheels to smear itself all over bonnet, cab and side doors. It is then inevitable that the slurry migrates onto the driver's clothing, as he dismounts to open or close gates, and fairly shortly afterwards the slurry is on the steering wheel and controls. Scraping is an unpleasant job, consuming much more time than it ought as a result. Very often the tractor is rather elderly and not the best starter, the better tractors being reserved for field work. For all these reasons, the half-hour job of scraping extends to over an hour in the less well laid out buildings.

Automatic scrapers

With the cost of labour increasing each year, scraping slurry is not the nil cost operation many believe. Added to this is the interest by more and more dairy farmers in scraping twice a day to maintain cleaner floor conditions in cubicles. The aim is to keep cows cleaner as well as the cubicle beds since clean beds are less susceptible to disease build up.

Thus there is at present a growing interest in automatic scrapers in cow accommodation. Their use is quite widespread in Europe, and it is strange that UK farmers should be so different. Perhaps it was the unreliability of the early designs tried in the sixties which got the whole concept a bad name. There was also the capital cost which seemed high in days when labour was cheap and cows cost well under £100.

Modern scrapers cost around £25–£30 per cow place (1979) and so are much cheaper in real terms than ever before. Better knowledge of metallurgy has resulted in better wear life and new designs have eased the installation problems considerably. Whilst designs differ in detail, a scraper installation will consist of a drag chain running along the cubicle passageways, in which operate one or more scraper boards which are pulled by the drag chain. This itself is powered from a small electric motor and gear box to give a speed of scraping around 3 metres per minute. A 1 kW motor and gear box can power a scraper with 200 m of chain, enough for 150–180 cows depending on layout. Usually one scraper is in work whilst the other is returning empty to its starting point and passageways of unequal length can be cleaned. Widths of scraper up to 3·5 m are available and the scraped manure can be pushed into a collection pit, sump or channel for removal at the gable end or at the middle of the building. (Plate 12).

Because of the slow rate of moving of the scrapers, there is little likelihood of injuring an animal; should a half-tonne cow stand immovable an overload clutch in the gearbox will prevent damage to the drag chain mechanism. Time of scraping can be controlled by a time clock to take place several times a day and so ensure clean floors with the minimum of supervision necessary from the cowman.

Both of these aspects are undoubtedly attractive, and taken together with the rising cost of tractor scraping, form a strong case for considering automatic scraping in livestock buildings. A benefit difficult to quantify but put forward by a number of thinking farmers, in favour of automatic scraping, is the avoidance of the 'disturbance factor' that an operator, who must interrupt one task to carry out a second job for half hour or so, very often does not return to the first task at the same workrate or intensity as he left it. The significance given to this case is very much a subjective decision but it has nonetheless a real effect—we all work better if undivided attention can be given to the job in hand.

- There are two other scraping mechanisms available which should be mentioned. In one case, a straight bar reciprocates forwards and backwards along the length of the channel or passage to be scraped. Fixed to the bar at intervals and able to pivot about the fixing point are ploughs or paddles (resembling a short bar in fact) long enough to span the width to be scraped. When the long drive bar moves forward, the paddles extend across the channel and sweep forward, retracting alongside the drive bar as it reverses. The movements are not rapid, about five or six per minute.

The second type of mechanism is a heavy-weight chain carrying at intervals short bars located at right angles and extending across the channel being swept. Bends can be accommodated using locating pulleys in the corners, and the conveyor can rise in height if there is a need to scrape manure up a ramp to the unloading point.

The long-term reliability of the various mechanisms is yet to be shown under UK conditions, so it is too early to pass judgment. Finally, where long straw bedding is used, it is important to check with the manufacturer that his equipment will deal with long straw: some designs really require bedding materials to be chopped.

Handling slurry

There is a vast array of equipment for this task and space does not allow comprehensive comment here. The types available fall into classes of pumps, augers, elevators and a recently announced pneumatic system.

Pumps

There are nowadays three main types used for pumping sludges and slurries comprising diaphragm, scroll and stator and centrifugal pumps. A decade ago it was not uncommon to find piston pumps fitted where organic irrigation was practised because they could generate high pressures to cope with long distance pipe runs. They have almost all expired now and are too costly, by a factor of three or more, to instal new.

The civil engineering world uses diaphragm pumps a great deal to pump sludges and semi-liquids. Small stones can be pumped, too, but long fibrous material present in the liquid will defeat the pump, mainly by lodging around valves, etc. This is not a wholly fair criticism of the pump since the presence of straw and other materials is the root problem in all farm waste liquid handling. Like piston pumps, diaphragm models can generate high pressures if needed but they are seldom found permanently installed in agriculture. They are sometimes hired for specific jobs but seldom for slurry because of the ready availability of pumps for slurry handling.

The third type of positive displacement pump is the scroll and stator variety, best likened to a corkscrew inside a closely fitting tube. Turning the corkscrew at moderate speeds gives the required pumping effect. As manufactured, these pumps have a stainless or other high quality steel rotor and the stator is rubber. Having no valves, they can deal with some fibre but if this is overdone and a blockage results, such is the pressure capability of the pump that pigs' hairs and husks in pig slurry can be compressed against a blockage to resemble damp chipboard. Available in sizes from less than 300 mm in length to over 1 m and a range of diameters, they can be powered by electric or other motor or by tractor pto shaft. Two-stage pumps are available where high pressure is required; for example, 11 l/s (40 m^3/h) flow at up to 10-bar head would need 18–20 kW shaft power. This type of pump is also fitted to some slurry tankers.

From the operational point of view, the outstanding precaution is to ensure that the pump *never runs dry*. If it does, the rubber/steel contact quickly heats and the rubber picks up on the shaft which then removes pieces out of the stator thus ruining it.

Centrifugal pump. The principle of the centrifugal pump is well known: a revolving plate carrying radial vanes or flights throws liquid from the centre to the periphery. The inlet supply is taken to the centre of the pump, and by shaping the casing around the impeller, outgoing liquid is collected in the supply pipe connection. A deal of experimental work goes into getting the most efficient shape of impeller and flight, and this type of pump has probably seen the greatest development to adapt it to slurry use. There are a great many models on the market mainly for tractor attachment and pto drive, (Plate 13). Varying lengths of pump are available so as to be able to reach to the bottom of slurry pits, and some models can be fitted with dual drive, *i.e.*, by pto shaft or electric motor.

Of course, this type of pump would be just as vulnerable to wrapping and blocking with wads of straw as any other. Manufacturers have spent considerable effort to design chopping devices integral or attached to the centrifugal rotor. The difficulty has been to achieve the necessary ability to chop the wide variety of materials in agricultural slurries and maintain this ability for long periods of arduous service. Hence, cutters of hardened and special steels are used.

No matter how good the design, the unit will fail prematurely if subjected to abuse, the chief of which is to allow into the slurry foreign objects like stones and metal which can blunt or fracture cutters, or plastic twine which can wrap around bearings. When inserted into a slurry pit, the pump should be positioned about 300 mm above the floor and preferably on the opposite side to the slurry inlet point. In this way the likelihood of picking up stones is avoided, presuming that a regular inspection of the emptied pit is made and stones and debris shovelled out. The amount of small stones brought in by a 100-cow herd can be truly surprising and regular checks are vital.

These pumps can deal with quite thick slurries and are used to macerate manure before loading into storage tanks or for direct loading of slurry tankers. When pumping thick slurry, such a pump can be a very full load for a 35 kW tractor on the pto, yet power requirements may fall to one-fifth when handling thin slurry. Given adequate power, throughputs of 100 l/s are possible and a five or six cubic metre tanker can be loaded in little over a minute. Although not able to generate the high delivery pressures of some other pumps, the centrifugal chopper mixer slurry pump can supply sufficient pressure and volume to power the jetting kits offered by most manufacturers of above-ground slurry stores.

Pump-mounting arrangements. As mentioned earlier, the standard design is for tractor-mounting, but optional support frames are available so that the pump can be left permanently mounted in the pit, requiring only the attachment of the tractor pto shaft to commence work. Sometimes the dairy scraper tractor is small and light and not man enough for the slurry pump. On other farms, the inconvenience of finding the tractor, positioning it and coupling up the power shaft is such that an electric motor is fitted. Whilst it is accepted that the cost of the motor is not a small item, many farmers are surprised at the wiring connection costs. Motors ranging from 8 to 20 kW may be used, but low power motors may often disappoint with their performance and are a very questionable decision.

Farmers are often concerned at the size of electric motor suggested by the installer, forgetting that their tractor pto can supply twice the power. The need for high horsepower is because the pumps are large: so 'why don't "they" make smaller pumps?' is a frequent question, especially where a mechanical solid/liquid separator has to be fed with only a moderate rate of slurry. The situation is that under practical working conditions, pumps mainly fail due to blockage of the inlet by lumps or wads of bedding. The way to minimise this is to use as large an inlet as possible, so presenting the maximum area of an approaching blockage to the chopper teeth. This means an inherently large rotor or, conversely, below a certain approximate size the inlet aperture can be too easily bridged by quite moderate obstructions. Thus too small a pump is a liability as it blocks so easily. In general terms, large impeller means large pump means large motor requirement.

In addition to the popular tractor-driven types of pump, there are also a few electric submersible designs fitted with cutters. These cost about half as much and can deal with a reasonable straw content in the slurry, especially if chopped straw or sawdust bedding is in use. These pumps are heavy and although they can be lifted and lowered into the slurry pit by tractor front loader, it is well worth installing sheerlegs and a block and tackle to handle the pump. Without this provision, life being the struggle it is, when the pump needs lifting to unblock it, the front loader tractor is sure to be elsewhere.

Before leaving the topic of pumps, it is worth reiterating that too much fibre in the slurry is the prime cause of failure, and so the lesson is to take steps to eliminate or at least absolutely minimise the amount of fibre. Provided slurry will run to a pump inlet point, the pump can normally manage to transmit the manure at least the short distance up and into a tanker from a pit, albeit at a reduced volume rate. The addition of water will help, and it is a management decision whether the extra handling caused by adding water is worth the higher pumping rate achieved. When pumping from a collection pit or sump, it is almost always worth the effort to homogenise the pit contents by a period of recirculation before extracting manure.

Where slurry must be pumped through a pipeline without fear of blockage, then dry matter should be reduced to about 3 per cent, which means a dilution of 2 or 3 : 1. Pipelines should be laid as straight as possible, avoiding dips in the general slope of the pipe run. The line should be flushed out with clean water after use to avoid corrosion of aluminium pipes, where used, and prevent any pockets of sludge settlement. It is a sound idea to arrange for the line to be self draining to avoid risk of frost damage.

The liquor from a separator machine may reach six per cent in dry matter, yet liquor can be pumped satisfactorily through small bore (38 mm) plastic pipe long distances. The explanation here is that the separator not only removes long fibre, but also the particle size distribution is altered too (see page 152).

Augers
Provided long straw is absent, a 100 or 150 mm diameter grain auger can elevate slurry in an emergency. The trouble is the corrosion and even with a water rinse after use, auger life is not long. Throughputs of 5 l/s were measured on one occasion accompanied by much clattering of the 100 mm auger in its tube.

Purpose-built augers for slurry handling are nowadays usually confined to poultry units for the conveying of battery cage droppings. Auger diameters of 200–300 mm seem to be common for horizontal transport at modest speeds of rotation, 100 revolutions per minute. Greater rpm are used for the last section of the auger used to elevate the droppings for deposition in a farm trailer.

Performance figures are hard to come by and can be misleading because of the variability in dry matter from farm to farm (leaking drinkers). In practice it is common to find a hosepipe inserted in the loading end of the auger to lubricate the slurry. The water reduces the dry matter of the droppings from 23–25 per cent to 17 or 18 per cent and greatly speeds up throughput, *e.g.*, 5 tonnes per hour through 30 m of 300 mm auger elevated into trailers using two electric motors totalling 5 kW. About every two years the ravages of corrosion are made good with the welding torch and new bearings, etc.

Elevators
The problems of pump blockage have been mentioned ad nauseam and testimony to the reality of the practical problem lies in the introduction by a major machinery manufacturer of a slurry elevator during 1977. The particular need was to ensure a steady feed to a mechanical slurry separator. A neat vertical plastic box section casing encloses a malleable iron chain carrying flights of reinforced rubber. Stainless steel shafts and Tufnol bearings further reduce susceptibility to corrosion. Variable speed drive is fitted to control throughput, which can exceed 6 l/s (22 m^3/h). The significance of this is that the power required is as little as 0·75 kW, a fraction of that required for any pump system. Costing just over £1,500 (1979) the elevator is cheaper than many pump systems.

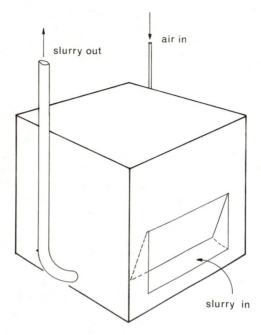

Fig. 17. Pneumatic 'letter box' slurry pump. Submerged in slurry, the box fills via the large flap valve. Air is then admitted to close the flap and force slurry up the pipe.

Pneumatic system

A pump powered by compressed air was developed during 1978 on a Lancashire farm. Interestingly, the pump idea evolved as an optional extra to add to a system using compressed air to agitate the contents of an above-ground slurry store (see page 98). Using air at 6 bar from a 2·25 kW compressor, slurry can be lifted 7 m at 55 m³/h.

The 'pump' (fig. 17) comprises a 600 mm steel box having an inlet like a letter-box (500 mm × 250 mm) in one side submerged in a pit. Slurry fills the box and air pressurises the box closing the one-way flap. As pressure continues to increase, slurry is forced up the 100 mm delivery pipe to the spreader or storage vessel. As the last of the 227 litres of slurry is discharged, air pressure drops rapidly in the pump chamber and this fall is detected by a pressure sensor which opens two solenoid valves to divert the compressor output to atmosphere. This allows more slurry to enter the box, and after an adjustable delay of only a few seconds the solenoid valves close, compressed air is allowed back into the pump chamber and the cycle continues about four times a minute.

The attractions of the system are its simplicity, robust nature and that it uses a fairly cheap low power source which typically would cost under £10 per year to pump slurry from a 100-cow herd. Drawbacks are that there is no chopping mechanism, and the leisurely rate of pumping when loading tankers for spreading, e.g., a 5 m^3 tanker would take almost six minutes to fill. This last objection could be met by filling an overhead tank during the spreaders absence.

HANDLING TO THE FIELD

A pre-requisite of successful unloading of a slurry store is that the contents should be thoroughly mixed *before* unloading begins. This has been stressed earlier concerning the management of stores. Earth-banked lagoons or very wet compounds will usually have settled out into their respective layers of top crust, bottom sludge, and so on, and require mixing before emptying.

One method is to position a tractor-driven high volume slurry pump at a convenient position on the bank and using maximum tractor power, jet the crusted slurry to mix it, beginning next to the pump and then outwards in increasing circles. With very wet slurry, this may be sufficient but it is time-consuming as many positions round the bank are necessary and it is notoriously difficult to mix in the bottom sludge. If too much of the liquid is removed, it becomes impossible to reach much of the sludge at the bottom as the pump volume delivered falls off as the slurry effectively thickens.

A much better method is to use a tractor-driven lagoon agitator. A tube 4–6 m long carries a propeller of three or four blades some 600 mm or so diameter. The tube contains a shaft transmitting pto drive from a tractor to the propeller. The whole agitator is hitched on the tractor three-point linkage. These devices, costing £800–£1,300 depending on length, can effectively mix lagoons and chop up islands of floating crust containing shrubs, weeds, etc.

The tractor is positioned at a few places along the bank as lagoon shape and size dictates, the propeller lowered into the slurry and the pto drive engaged. The whole lagoon should be mixed *before* unloading begins and up to two working days may be necessary to achieve this. It is tempting to set up the tractor and leave it unattended, but it is much better for a member of the farm staff to work close by to keep things under observation. It is no bad thing to make sure the tractor handbrake is on and to chock the wheels too lest the vibration surreptitiously walks the tractor into the soup.

TANKERS

There are something like thirty manufacturers offering over 160 different models of slurry tanker ranging in size to more than 13 m^3 and so it would be a hopeless task to try to mention all the various design aspects here. However, a few comments might be useful.

Types
(i) There are open topped tankers, filled using an external pump and discharged usually via a rotating spinner at the rear. Drive from the tractor pto is carried by a shaft passing through the tank to the rear spinner. This shaft may also carry a mechanical mixer or auger to carry and force the slurry into the spinner. Where a stone trap is fitted, it will be at the rear in front of the spinner. These machines are cheaper by £300 to £1,100 (1979) than the equivalent sized machines fitted with vacuum pumps and can handle the thicker slurries (12 per cent).

One maker has produced a self-loading tanker. The unloading spinner distributor is reversed in direction and so beats slurry into the tanker. This is effected by constructing a sump or depression in the lagoon. The tanker is reversed into the sump until the slurry level reaches the spinner which then commences the filling operation. The driver does not have to leave his seat and so the design appeals to owner drivers.

(ii) True tankers fitted with an air pump which can be used to evacuate the tank to suck in a load. Change-over valves then allow the pump exhaust to be used to pressurise the tank up to 2·76 bar and so expel the load through a mechanically or hydraulically operated valve at the rear of the machine. Tank contents are agitated by blowing compressed air into pipes in the bottom of the tank.

These vacuum fill pressure discharge tankers can handle slurry of 10–12 per cent dry matter and may be used for agitating a pit or tank with an air line.

(iii) There are also a smaller number of tankers fitted with positive displacement liquid pumps used for filling and emptying liquids and slurries. Depending on the pump size relative to tank volume, they can fill faster than vacuum-fill models. Like the pressure discharge types, they can be used for store agitation and can discharge from a hard road onto the field via a pipeline up to 220 m long, feeding to a rain gun.

Sizes

2·2 m^3 to 13·6 m^3 covers the range, most popular sizes lying between 4·5 m^3 and 6·5 m^3. There are two common mistakes made in choosing tank size: too small or too large. Too small a tank is often bought as a result of a misplaced zeal to 'keep the cost down'. The cost of a machine never solely resides in its purchase cost and some study of its working performance will often warrant paying for a larger model.

The oversized tanker frequently arrives on a farm to replace a model which has proved too small for the farm's needs. It is a natural reaction, then, to buy a new tanker much, much bigger and only after the first journey to realise that the farm lacks a tractor big enough to handle the fully-laden tanker safely. Even a 5 m^3 tanker, which is not overlarge in payload, will weigh 1·5–1·6 tonnes empty. 5 m^3 of slurry will add another 5 tonnes to give an all-up weight of over 6·5 tonnes which, for a medium horsepower tractor weighing, say, 3 tonnes, is more than a handful on slopes or at transport speeds on the road. As the payload capacity increases, so must the laden weight of the tanker so that a 9 m^3 model will weigh a total of nearly 12 tonnes.

Wheels and brakes

Two aspects follow the size question: tyre equipment and brakes. It is commercial practice to market a machine at its most attractive price, and a common economy is to fit the smallest tyre size which will carry the load. What the farmer must realise is that this might be a very different capability from what road and track conditions on his farm dictate in the way of tyre equipment. Good service life is not just longevity of the tyre but freedom from punctures too. These are important on the farm and so buying extra quality, usually thicker casings (referred to as having a greater number of plys in the carcase) is always worthwhile.

Again, on commercial pricing grounds, braking systems are often not part of the standard equipment on the tanker. There are the strongest possible reasons on account of safety for buying a tanker fitted with brakes and few would argue with that. The word of warning is to ensure the braking system offered is compatible with the tractors likely to be used and that it is sensitive enough to brake the wheels without locking them when the tank is empty. If locking does occur, tyre life is much reduced.

Where tandem axles are fitted, high ply tyres are necessary to stop the tyre being pulled partly off the wheel rim when turning tightly on hard surfaces. A few tankers are offered with a hydraulic system to raise the front wheel of the tandem pair, so easing this problem. The tandem axle is fitted to give good flotation over fields and to reduce ground pressures within acceptable limits to minimise rutting of soil surfaces.

Discharge pattern

There are two considerations here: evenness of application and pattern. A basic assumption is that manure spread onto fields or crops must be spread fairly evenly to avoid problems of varying crop growth and maturity. Looking at the discharge pattern from many tankers and the deposition on the soil surface, it is plain that application is far from even. There is no record of any investigations having been done into this and, if so, *it is probably the only* topic in agriculture which has not been researched! The reasons for the lack of interest are not difficult to imagine; apart from the technical difficulties, there is the little matter of getting rid of the smell from the investigators at the end of the day.

Spreading pattern is becoming much more important nowadays. It was common to find tankers discharging their load in a halo of spray from the rear, sometimes reaching 10 m into the air. It was a selling point as a greater bout width was covered. The current problem is to achieve a reasonably even spread yet at the same time project the slurry as little as possible into the air, to minimise the problem of odour generation. Once the severity of the odour problem began to bite home in agriculture, manufacturers were fairly quick to re-design spreading devices to obtain a low trajectory for the slurry. This is now a most important requirement for any agricultural buyer.

Other details

The tanker should be fitted with a stone trap which should be easy to get at and have a quick release unloading hatch. Stone traps which are a fiddle to empty are a liability as they are not regularly emptied which defeats their purpose.

Tanks need cleaning out and maintenance. Whilst a manhole access door is cheap, it is not everyone who will relish crawling around inside a slurry tanker needing a clean out! Far better to choose a design where the complete back plate can be hinged open by releasing several clips.

All hose attachments should be via *convenient* quick-action clips, and there should be plenty of storage space on the tanker for suction hose—one length is so often not enough to reach into buildings. It is no solution to have to drape extra hose over the tanker for transport from one site to another.

Where much road work is undertaken, it is sensible to equip the tanker with lights and direction indicators if only to reduce the nervous stress of the tractor driver who may not have a clear rear view in making turns, if not to help ward off motorists.

A tanker has other uses too: filled with water it makes a good pressure washer for cleaning off equipment, etc; it is also a useful water tender for use against fires or as a precaution to have handy when burning straw. Tankers have on occasions been used as a standby source of vacuum when power cuts or other failures have crippled the normal milking machine vacuum pump, even though this is not an ideal practice from a hygiene point of view.

Power requirements

This detail can be quite deceptive in that the shaft power to run the auger spinner of an open-topped tanker may amount to 25 kW and less for a vacuum pump, but the size of tractor suggested by the manufacturer may be 75 kW. The disparity is because of the need for stability of the whole outfit during transport and work.

Work rate

An open-topped tanker can usually be filled by a slurry pump in just over a minute. A vacuum tanker may take three or five minutes to fill, very large tankers taking longer. Most tankers can be emptied within five minutes in the field, the limit being set not so much by the tanker as the need to work the field evenly and to make turns out of work.

In terms of calculating the overall performance of a tanker during spreading, the greatest determinant is usually the journey time which so often occupies most of the working day, the filling and spreading being quite a small proportion. It is a common mistake to add together a couple of minutes' filling time and twice as long to empty plus twice the journey time and divide this total number of minutes into the hour to produce an hourly rate of work. This totally overlooks the reality of the job which is boring, messy, stinking and generally unpleasant. It takes a fastidious driver to prevent slurry migrating from the suction hose onto his clothing, his tractor seat and finally onto the steering wheel. After a day's spreading, getting rid of the colour and smell of slurry from hands, arms, etc is a formidable task. These conditions are seen reflected in the rather measured pace with which the job is tackled hour in, day out—which may be a slower rate than when the boss or visitors are watching. Numerous timed studies of farm operations have rarely shown a sustained work rate over three loads in the hour where journey distance is further than the next field or two. For planning purposes there is likely to be less disappointment if a rate of 2·5–2·8 loads per hour is used.

Contractors will normally exceed this quite a bit, but their drivers are often on piece rates of earning and seem to be equipped with grand prix tractors allied to a rather cavalier attitude to equipment and sometimes the Highway Code.

Costs

Autumn 1979 prices suggest a figure around £3,500 for a 5 m^3 pressure fill and discharge tanker and around £1,000 less for an open-top mechanical discharge model. In the jumbo size, prices approach £7,000 for a 14 m^3 vacuum tanker and just over £6,000 for an open-topped version. It should be said that most tankers are priced without many of the so-called extras which the prudent purchaser may consider quite necessary. One particular 5·5 m^3 model, priced at £3,420 ex-works, totalled £3,975 delivered equipped with stronger wheel gear, mudguards, lights and brakes, so great care is needed if comparing prices.

Calculating the running cost of a machine is always ripe for argument amongst economists. Some suggest that, for instance, tractor running costs should be treated as a farm overhead since a large element of tractor time is of a general nature and cannot be allotted very precisely to the various farm enterprises. However, for a job like manure handling or spreading, the tractor cost is recognisably a direct one. Table 17 shows the costs for three popular sizes of tractor.

As explained elsewhere, the weight of tractor needed for tanker work is determined by the load to be braked and controlled and

TABLE 17. Guide to Tractor Operating Costs

Power in kW	45	60	75
New cost £	7,500	9,000	13,000
Annual usage (1)	-----	1,000 hours	-----
Economic life	-----	7 years	-----
Terminal value £	2,500	3,000	4,000
Interest	-----	18 per cent	-----
ANNUAL COSTS			
Annual cash flow of investment (2) £	1,762	2,114	2,081
Fuel:			
Average usage l/h (3)	(4·1)	(5·0)	(12·0)
Cost @ 11p/l £	451	550	1,320
Lubricating Oil:			
Annual usage l	(90)	(110)	(135)
Cost @ 42p/l £	38	46	57
Repairs and maintenance			
@ 8% new cost £	600	720	1,040
Tax and insurance £	75	94	125
TOTAL	2,926	3,524	5,623
Cost per hour £	2·93	3·52	5·62
Driver per hour £	1·60	1·60	1·60
Total operating cost per hour £	4·53	5·12	7·22

Footnotes:
1. Many tractors achieve only 750 hours per annum.
2. Annual cash flow is calculated using a standard programme which takes account of price, interest, economic life and terminal or scrap value of the machine.
3. Fuel consumption figures look low but they are *average* figures from ADAS Farm Mechanisation Study No. 21 which recorded tractor and fuel use over a complete year.

not by the horsepower required, since pto requirements are easily met by the 45 kW class of tractor. After the initial cost price of a machine, the factors influencing its cost are: the interest on borrowed money, the annual usage (hence economic life) and the value the machine makes when sold. This last element is the most difficult to assess and the values used in Table 17 must inevitably be estimates.

In determining the running cost of a tanker, much depends on its annual hours of use. Because of the corrosive nature of slurry,

TABLE 18. Slurry Tanker—Running Costs

Capacity Price	5m^3 £3,500		9m^3 £5,100	
Annual use in hours	100	200	100	200
Life in years	7	5	7	5
Cost per annum:				
Depreciation*	902	1,091	1,322	1,603
Repairs and maintenance	210	350	306	510
TOTAL	1,112	1,441	1,628	2,113
Cost per hour	£11·12	£7·20	£16·28	£10·57
Cost per m^3/h**	£0·89	£0·58	£0·80	£0·52

* Interest at 18%; terminal value £200
** Estimating 2·5 loads per hour

even a lightly used tanker will have a life expectancy not so very much longer than a hard-worked but well-maintained example. Table 18 sets out a method of costing a tanker of 5 m^3 and 9 m^3 capacity.

Some tankers are used for more than 200 hours a year and this will reduce their hourly cost a little. The precise cost of tankering slurry will vary with size, cost and usage of both tanker and tractor. From Tables 17 and 18, the range of costs lies between £11·73 and £23·50 per hour depending upon the combination of tractor and tanker size used.

Another way of looking at spreading costs from these tables is to presume 2·5 loads per hour in which case, cost per m^3 spread ranges from £1·47 to 70p as set out in Table 19.

TABLE 19. Cost in £/m^3 of Spreading Slurry by Tankers
(assuming 2·5 loads per working hour)

| Tractor Size
kW | Tanker size and annual use | | | |
| | 5 m^3 | | 9 m^3 | |
	100 h	200	100 h	200 h
45	1·25	0·94	Tractor unsuitable	
60	1·30	0·99	0·95	0·70
75	1·47	1·15	1·04	0·79

Soil Injection of Slurry

An alternative method of applying slurry to land is to inject the manure below the upper soil layer via a tined implement and this was certainly practised before 1870. The over-riding advantage of this is the large measure of control of smell achieved. Such is the success of the technique that in some areas in Europe, slurry may only be applied to land by injection. Although one local authority in UK became very attracted by the idea and could well have tried a similar approach in their area by instituting suitable legislation, the barrage of objection by agricultural interests won the day, fortunately.

Two other advantages claimed for soil injection of manure is that since the slurry is placed typically 75–100 mm below the soil surface by the tine and immediately covered up by the displaced soil's falling back over the slot cut by the tine, there is no opportunity for loss of N by volatilisation. This contrasts markedly with normal tanker application where not only is slurry left on top of the field, but manure is airborne during the spreading process—both aspects encouraging loss of any volatile N.

The remaining advantage claimed is that run-off of liquid manure from the field surface into ditches is prevented. This in practice is true only if the operator is very conscientious in controlling the application rate to meet crop nutrient requirements, *i.e.* avoids excessive rates. In addition, if the field surface slopes even a small amount the slurry tends to run back along the slot and overflow onto the soil surface in the troughs: so run-off can occur depending upon the application rate, the absorbency of the soil and the amount of slope in the field.

On the debit side of injection lie eminently practical considerations. There is the agronomic penalty in that, whilst injection into a stubble or cultivated soil is no problem, tines ripping a grass sward continuously at intervals of half a metre leave a terrible mess and undoubtedly reduce the field's productivity. Grazing animals can utilise the grass (and will voluntarily return to grazing almost immediately), but grass mowers can be in difficulty with the rough surface. It is incidentally not always a ready solution to use a roller immediately after injection since the weight squeezes out manure onto the surface. If the roller is put in after a day or two to allow the slurry to be absorbed, it is common to find that the divots of turf have dried and do not fit neatly back into the slot.

Perhaps the greatest disadvantage of all is the large power requirement to pull a tanker equipped with an injection frame. Although no specific tests have been carried out in the UK to measure power requirement, American experience reports 5 to 12 drawbar horsepower per tine necessary. With four or five tines at work and the seven tonnes or more of tanker to pull, the job very often needs a four-wheel drive tractor of around 75 kW. With such weight and power at work, if field conditions are overdamp, rutting can occur, adding further to the damage sustained by the crop. These considerations can combine to greatly limit the periods in the year when soil injection can be fitted into the working farm timetable.

Equipment

First of all, the tine itself. There are a variety of designs but essentially a pointed soil-engaging tine opens a slot in the soil and slurry is introduced into the slot behind the point usually via a shaped pipe. In turn this is fed from a distribution point or manifold receiving slurry under pressure from the supplying tanker. Each tine may be preceded by a disc coulter to allow a neat cut into the soil and two to five tines may be used capable of injecting to depths of 250 mm.

Because slurry contains so much fibrous material, blockage of manifold and pipes can be frequent. Designers have tried to cater for this by quick-release arrangements for pipe ends and on hatches, to allow rapid access for removing blockages, etc. Even so, soil injection can be a frustrating and slow job if the wrong sort of slurry is being used because of the frequent stops. The ideal material is the separated liquor from a mechanical solid/liquid slurry separator as fibre is removed in the separation process.

Injector arrangements

There are three arrangements used for injectors:

(i) A slurry tanker carrying a toolbar injector at the rear (Plate 18). This makes the injector easy to attach but it is difficult for the driver to see the work and to look for blockages. In addition, when the tanker is empty, the weight downwards of the rear frame tends to lift up the tanker drawbar resulting in reduced grip of the tractor rear wheels and lots of chatter at the tractor drawbar which will displace an unsecured drawbar pin.

(ii) A more compact unit is available where the injector is carried on the tractor three-point linkage in such a way as to allow the tanker drawbar to be hooked up in the normal way. This arrangement gives maximum traction to the tractor, but only two tines can be accommodated to avoid fouling the tanker drawbar during running.

(iii) As a broad generalisation, predominantly livestock farms use only moderately sized tractors, yet injection requires high horse power. In Austria this potential contradiction was met by using one medium power tractor to operate the injector toolbar, which is then supplied with slurry through a large diameter hose resembling an elephant's trunk from a slurry tanker pulled alongside by a second medium power tractor.

This system really needs three or more tankers to keep the injector at work, which is the key element. To assemble a four-man gang is much more of a problem than to send off one man alone to spread manure and this is raised as a main objection to this Austrian system. However, it does work given cooperative endeavour from a few farms to make up a gang, perhaps sharing the ownership of the equipment, and avoids the very costly high horse power tractor otherwise necessary if a contractor service is not available.

Work rate

The tanker-mounted injector spends most of its time not working, *i.e.*, riding to and fro with the tanker from field to store and return. The forward speed whilst injecting slurry is not great, about 6 km/h and there are often frequent blockages to further slow down the output. Even where there are no stoppages at all, injection is quite a lot slower. For example, where 20 m^3/ha of slurry is to be applied using an ordinary 5 m^3 tanker at 6 km/h, then a typical time analysis might show 5 minutes to fill; 8 minutes journey to field; 3 minutes to empty (*i.e.*, 0·25 ha per tank load, spreading width, say, 10 m then distance to unload 2,500 m^2 ÷ 10 m = 250 m. If speed in work is 6 km/h or 6,000 ÷ 60 metres per minute = 100 m/min, time to unload is 250 m ÷ 100 m/min or 2·5 min) and finally 8 minutes to return to the store, giving a total elapsed time of 24 minutes.

The same load injected would increase the period for unloading (injecting) to 12·5 minutes (2,500 m^2 ÷ 2 metres spread width = 1,250 metres). At 6 km/h or 100 m/min this would take 12·5 minutes. Hence, total elapsed time would increase to 5 + 8 + 13 + 8 = 34 minutes, or an extra 42 per cent. In round terms, almost half as long again and that without the penalty of blockages.

Thus, although injection can be a very helpful smell control technique, it is not cheap—tines wear out and high power tractors are needed—and it notably slows down slurry application to land.

ORGANIC IRRIGATION OF SLURRY

The idea of using an existing irrigation system to apply manures to land looks attractive at first sight—fairly well automated equipment and two jobs in one (water and fertiliser). Such were the views twenty years or more ago when irrigation itself was taking off. However, only about 800 installations were ever declared in the agricultural census as being used for organic irrigation and this has now fallen to under 300 holdings.

More detailed consideration of operating organic irrigation makes clear the inherent restrictions of the technique. The generally accepted maximum dry matter for organic irrigation is five per cent, and many operators prefer to see three per cent dry matter to lessen the chance of blockage. At either level this implies considerable dilution of the slurry with water. This means extra costs for the water itself, extra costs for handling and pumping the increased volume, and it may well mean that increased storage capacity is needed to carry over manure for those periods when weather or unavailable fields prevents irrigation—again extra costs.

Although the rain guns are designed to accept some lumps without blocking by the use of rubber expandable nozzles, irrigation is not an easy bedfellow with the use of long straw bedding. Chopped straw is a must though most organic irrigation systems have a macerator or chopper unit ahead of the pump.

That organic irrigation became much less popular is borne out by the drop in the number of farm plants reported in the census. Any one or more of several reasons are likely to account for this including the following:
- Wet or badly drained land is unsuitable.
- The job is a very unpopular one with workers.

  About 12 m^3/ha is the advised limit for one application and this results in the rain gun needing to be moved frequently as the rate of delivery is usually high (large nozzles reduce blockages but result in high throughput). With a conscientious operator in charge, it is not unusual for the rain gun position to need changing every twenty minutes. With two guns to

look after, the operator is continuously engaged in moving pipes and guns, all of which are coated with dilute slurry. Even though clad in protective clothing, the operator very often ends up liberally coated too which is most unpleasant. To have to spend a day or more each week at this task is daunting and more operators are refusing.
- Distribution is uneven in windy weather.
- Frost can be a problem.
- Flushing out a pipeline after use can require quite a volume of water, which further wets the land.
- The system is costly to instal.
- Of increasing significance these days is the aerosol effect whereby airborne slurry droplets may be carried long distances and together with the unpleasant smell can be the cause of virulent complaint from domestic dwellings.

Currently there is some interest in mobile irrigators to ease the labour problem since one position of the unit can cover up to four hectares. The rain gun is mounted on a sledge which is gradually pulled in towards a large rotating hosereel carried on a wheeled chassis and positioned at a headland (Plate 20). The hosereel unit may be driven at variable speed to winch in the rain gun sledge using the 75 mm supply pipe to the gun as a winch wire. Water or slurry is supplied to the unit from an external pump, and power for winching may come from a tractor pto, a small engine on the chassis or by a water turbine driven by the pumped water (the turbine is not used on slurry work). One design incorporates an air compressor and air storage bottle so that after irrigation is completed, the pipework can be blown through to clear out the last slurry.

Along somewhat similar lines but smaller in size is a recently marketed low level slurry irrigator for separated liquor (Plate 27). This model is also supplied by an external pump, but the hosereel unit carries a wide spray boom fitted with simple jets to spray liquor evenly down onto the field surface. The reel and boom unit creeps along the field by winding up a previously laid out winch wire secured to a land anchor. About 2 hectares of land can be covered at a setting, the sprayed swath being 45 m wide and the length of run 426 m. This machine offers even application of what is really liquid manure and very little smell seems to be generated during its use which is a tremendous selling point.

In closing these few comments about organic irrigation, it should be said that the use of the system imposes a duty to use it correctly lest offence is given to the public. It is too easy to leave rain guns too long in one spot and seriously overdose an area of land. This could well result in polluting water supplies, which is an offence, and, worse still, may antagonise local people.

The greatest care should be taken to prevent aerosol or smell nuisance to the public as this too will cause antagonism. It must be clearly recognised that if the public are seriously disturbed by careless operation of rain guns, the offender may reap personal ostracism, the displeasure of the courts and could hasten the day when public disquiet over health aspects of organic irrigation boiled over—to the possible detriment of all users of the system, even to its prevention.

HANDLING FYM

Probably most solid manure is handled by tractor front-end loader fitted with appropriate forks or grabs. There are also a few three-point linkage rear-mounted versions too. Modern high capacity loaders can lift over a tonne of muck and can load at a rate bordering on a tonne a minute. Enthusiastic driving can achieve high loading rates but at some cost to the tractor transmission due to the constant clutch operation and gear changing necessary. A winter's use on this task will often result in some workshop attention being necessary for the tractor.

When buying a tractor which may spend a deal of its time on muck loader work, it is useful to check that the gear change allows convenient selection of forward or reverse—some models require a change of two levers instead of one, which is time-consuming and frustrating for the driver.

There is often a problem of rearwheel grip when clearing up an outside midden area. Fitting a heavy rear ballast weight is essential, and although it is a nuisance, changing rear wheels over so that the tyre treads run opposite to the usual direction can add usefully to rear wheel grip.

The three-point linkage rear-mounted muck forks do not have the height reach of front models. Better wheel grip is available because all the extra weight is over the rear wheels. It is still worth changing tyre direction as the need is for grip to get a good bite into the heap when the tractor is empty. The biggest objection to rear muck loaders is from the operator's point of view—the human neck lacks a universal joint and several hours work can be very unpleasant!

Slew loaders have been mentioned earlier and their high capacity of work, accurate control and reduced demand on the tractor transmission are all advantages to offset their capital cost (£2,000 onwards). Their output is very dependent upon the operator's skill, perhaps more so than with other equipment. Models are now available which can be mounted between the tractor and a towed spreader, so making a one-man unit for loading and spreading.

Next up the scale of costs comes the rough terrain fork-lift truck (RTFLT) in both 2- and 4-wheel-drive versions. Capable of traversing rough going, fitted with power steering and ergonomically-laid-out driver controls, these vehicles are rapidly being taken up by agriculture. Muck handling is one of their capabilities, and forks giving capacities of $1\frac{1}{2}$ tonnes at a bite are commonly used. Fitted with a top-acting grab cage as well, higher capacities are available but care is needed by the driver not to load spreaders with such large lumps; teasing out by use of the tilt and drop rams on the mast is necessary.

Crawler tractors fitted with large forks are particularly effective on rough or wet sites for field loading. For loading side spreaders, bucket size should be limited to about half a cubic yard (0·38 m^3); otherwise, with larger forks, spillage from the spreader during loading will be unacceptably high.

The dragline excavator once was the universal tool for drainage schemes and was sometimes to be found on muck handling, usually emptying lagoons or compounds. They are rarely used nowadays because of the high transport costs between sites. Very large buckets could be operated but half a cubic yard was large enough. With a 15 m jib and an operator skilled in throwing the bucket using the controls, a reach of 20 m was possible.

On an important point of safety during work with long reach loaders, it is a very necessary precaution to check for the presence of electricity supply lines and to ensure there is no possible chance of a flash-over onto a loader, apart from the obvious danger of actually touching the conductors. The inspection for this hazard should begin as far away as the turn-off from the public highway onto the farm road, into the yard or wherever unloading from the road transporter takes place and thence along the route the loader may take to reach its appointed working area. There are few second chances given by power lines to their victims, whether they be unfortunate or careless.

Work rates of FYM loaders

These obviously vary enormously depending upon the material being handled, the operator's skill and the conditions on site. However, as a guide Table 16 quotes data produced at the 1976 Power Farming Conference by Mr. D. A. Williamson of ADAS Mechanisation Department.

Transporting and Spreading FYM

For a generation, prior to 1962, a muckspreader was a flat-bed trailer with a combined chopping/distribution mechanism at the rear, the floor of the trailer carrying a slow-moving chain and slat conveyor. FYM loaded into the trailer could, when the spreading and conveying mechanism was engaged, be spread in a passably even swath out of the back of the machine. Power was taken from the trailer wheels, later from the tractor pto shaft, and rate of manure spread could be varied by adjusting the travelling speed of the floor conveyor. When filled by hand fork, the spreader performed well enough but once tractor front loaders appeared on the scene and the muck was then dropped into the trailer in lumps, the fragile nature of chains and drive components was revealed and it was not uncommon to have to hand unload a broken spreader to be able to rectify faults and breakages. Often poorly designed and unsuited to spreading slurry or sloppy manure, they were ripe for the near eclipse they suffered upon the arrival from the USA of the side delivery rotary spreader.

Side delivery rotary spreader

This is universally referred to as the rotaspreader, which is in fact the trade mark of Howard Rotavator Co Ltd.

This new design offered robust construction, simplicity of operation and mechanism and the ability to handle slurry of any consistency. The spreading mechanism is a pto-driven shaft running front to rear of the machine and carrying short lengths of chain. When the shaft is driven at between 260 and 300 rpm, the chains have sufficient centrifugal force to break up and discharge the material over the side of the machine. In work the end chains begin spreading first, and as they clear space, so more chains unwrap from around the shaft to begin speading. It is important not to overload the ends of the machine with manure as this prevents the first chains breaking free of the load. The side spreading gives good distribution to the edges of fields and the tractor driver can more easily view the work than with a rear spreader.

By changing the forward speed, the rate of spreading can be varied between 12 and 74 tonnes per hectare, the width of spread commonly being three metres. There are a host of different makes of the same idea now that the original patents have expired. Sizes vary too, but as an example one manufacturer lists three models of 2·5, 4 and 5·3 m^3 nominal cubic capacity, capable of carrying 1·4, 2·2 and 2·8 m^3 of slurry respectively. Suggested minimum tractor sizes required are 25, 30 and 40 kW at the pto shaft. This is because a fairly high starting torque is needed to establish the shaft speed necessary to extend the spreading chains. Power consumption is usually highest with slurry and, in an NIAE test of the smallest model, ranged between 17 and 30 kW. In the same test the actual weight of manure contained in a full load was 1·8 t of well-rotted manure but only 1·2 tonnes of long strawed muck.

These machines can also be fitted with a windrowing shield which allows silage to be carted and unloaded against an animal feed fence. There are also accessories available to narrow the usual width of spread and to carry more slurry without spillage, as well as a shield to prevent slurry being splashed forwards over the tractor.

One manufacturer produces an interesting hybrid rotary spreader. In addition to the usual flails for solid manure, the machine is fitted at the rear with a pto-driven slurry spinner. The two spreading mechanisms, flails and spinner, have individual drive shafts, with alternative pto connections available at the front of the machine. Another maker's design incorporates provision for the barrel body of the machine to be rotated 45 degrees to either side so as to contain the maximum amount of slurry. This also allows the body to be emptied completely and avoids the need for a lid to maximise the volume of slurry which can be carried.

Rear delivery spreaders

Very shortly after the marketing of the rotaspreader, new designs of the rear spreaders were produced (Plate 16) in the fight back for sales. Much heavier construction all round, especially in the drive mechanism and provision for handling slurry, overcame the problems mentioned earlier. Two inherent advantages of these machines were lower power requirement and easier filling, having a completely open top. Again referring to an NIAE test from

PLATE 15
A 10 m³ slurry tanker fitted with mudguards and lights (broken!) The quick-action hose attachment is particularly important. Note the twin large flotation tyres.
Author

PLATE 16
A slurry tanker at work on autumn stubble. The lower the spread trajectory, the better to minimise odour generation.
British Farmer & Stockbreeder

PLATE 17
Close-up view of the tines of a rear-mounted slurry injector. Alfa-Laval Ltd

PLATE 18
A slurry injector at work. With ground conditions just right, high quality work with little damage to the field can result. The driver cannot see the injector toolbar. Alfa-Laval Ltd

PLATE 19
An Austrian scene. A slurry injection frame is pulled by one tractor and a second pulls a slurry tanker in parallel. Slurry is transferred to the injector by the flexible hose.   Bauer Ltd (British and General Tube Co.)

PLATE 20
Becoming very popular for water irrigation (as here) but adaptable for spreading dilute slurry, the travelling reel type of irrigator. The tractor operates the pressure supply pump and also the reel rotation to slowly draw in the raingun.

Farrow Irrigation

PLATE 21
The gun mounted on its wheeled sledge.

Farrow Irrigation

PLATE 22
A side delivery rotary spreader at work.

Howard Rotavator Co. Ltd

PLATE 23
A rear delivery muckspreader which can also be fitted with a 'door' to retain liquid slurry during transport. The RTFLT loader can fill the spreader in two or three bites.

1966, with a net four tonne load, 11 to 18 kW pto power was adequate.

To enable slurry to be carried, an hydraulically-operated dam or door is available to fit in front of the rear beaters and to retain the liquid manure in the spreader body. The door is raised gradually when spreading commences. Approximate spreading width is two metres or just over and rates up to 45 t/ha.

Although both side and rear delivery machines are widely used by contractors, there are signs of increasing popularity of the wagon-type rear-delivery machines. This is possibly because being completely open topped, they are easier to fill fully and there are large capacity models, 7 tonnes payload and more, available.

General-purpose trailers

Where very heavy dressings of manure are required, it is the practice to sometimes use ordinary tipping trailers. High capacity models are often used to give large payloads, which can quickly be loaded using large buckets, forks or grabs. The loads are tipped to a pattern in the field, and at a convenient time the loader tractor will spend time in the field pushing the heaps out over the surface. Further spreading may be achieved by the use of tractor-mounted cultivators.

This is a somewhat rough and ready system perhaps, but it is probably practically acceptable where the aim is to apply generous dressings to a soil low in organic matter status rather than an exercise in the precise fertilising of a crop. Because of the damage done to the field surface, the method is usually reserved for autumn stubbles or other crop aftermaths.

CALCULATING APPLICATION RATES OF FYM

On many farms decades of experience mean that farmer and worker know that certain fields usually have a particular number of loads and there seldom need by any calculation of the required spreading rate. However, a method of making the calculation may be useful:
1. Decide required application rate *e.g.*, 75 t/ha.
2. Determine or estimate average load weight in spreader, *e.g.*, 3 t.
3. Divide No 1 by No 2 to discover number of loads needed per hectare, *i.e.*, 75 ÷ 3 = 25 loads/ha.
4. Divide 10,000 by No 3 to obtain number of m^3 to be covered

by one load (there are 10,000 m² in 1 ha) *i.e.* 10,000 ÷ 25 = 400 m² per load.
5. Divide No 4 by average width of spread to find distance to be run with each load. For example with a 2 m width of spread: 400 ÷ 2 = 200 m run.

It is not an onerous task to pace out quickly the distance covered with the first load and to try a gear higher or lower with the next load. This will rapidly show the gear to use to achieve the required application rate or close to it.

It is possible that the required application rate is higher than the machine can achieve in one pass. In this case the solution is to work to half of the target rate and apply two coats to the field working at right angles to the first pass with the second.

Having carried out this check, it should be realised that much depends on the average load weight remaining constant. Where patches of 'strawy muck' are encountered during loading and spreading, a higher application can be used to compensate either by using a lower gear or spreading a narrower swath by driving closer than usual to the previous bout. This is perhaps not very scientific in accuracy but in practice is probably a reasonable working arrangement.

REFERENCE

LILLYWHITE, M. S. *et al*, The Flow of Pig Slurry in Open Channels. Note 102/76, Building Research Station (1976).

Chapter 7
SEPARATION OF SLURRY

THAT SLURRY is a mixture of faeces, urine and other materials to produce a semi-liquid of varying dry matter has been amply discussed earlier. Pumping a semi-liquid is difficult, blockages are likely and when left in store for a time, slurry separates out into sludge, a liquid fraction and perhaps a crust of the fibrous material. Sampling such a store is difficult, as will be the emptying due to the problem of re-incorporating the various fractions to allow handling hydraulically for application to the land.

Almost all of these handling difficulties can be eliminated if the slurry is treated to remove the larger particles, leaving the truly liquid fraction. It is this process which is referred to as 'separation' and may be brought about by a form of slow filtration or at a much more rapid rate by a machine.

The best example of the slow filtration method is the strainer box or chimney mentioned earlier in chapter 5 when discussing the management of storage compounds. Similarly, above-ground sleeper compounds, constructed of walls gapped vertically or horizontally, slowly filter out the liquid leaving behind the fibre, etc. Again, as mentioned previously, the separated liquid (SL) resulting can be pumped easily long distances through small bore piping with quite moderately sized pumps of any type.

By using a machine for separation, the process can be carried out at a much higher rate, producing under all conditions a stackable fibrous solid and a free-flowing liquid. At present, the uptake of separators by agriculture is going ahead slowly, largely on account of the costs incurred. A separator listed at £4,000 in a manufacturer's catalogue (1979) is only part of the cost. By the time the machine has been installed on its gantry; equipped with (a) a feed pump, (b) piping to remove SL to storage, (c) perhaps a reception pit and chopper pump and (d) necessary electrical wiring and general safety requirements, the all-in cost may have risen two and a half times or even more.

Discussion of the advantages of separation usually focuses on the improvements which result in the handling of the slurry, and these are real enough. However, following on from the work being done by Dr B. F. Pain at the NIRD, there is an equally strong case for separation on the basis of the agronomic benefits compared with the use of whole slurry on crops—which aspect is dealt with later.

PRODUCTS OF SEPARATION

The process produces separated liquor (SL) and separated fibre (SF).

Separated fibre (SF)
This comprises the larger particles in the manure together with pieces of wasted food and bedding, resembling a friable brown material with much less smell than the original manure. SF will usually amount to about one-fifth of the original slurry volume.

In general terms, SF appears to stack satisfactorily provided that its moisture content is not less than 16 per cent for cow manure and 20 per cent for pig manure. At these levels, there is very little seepage of liquid from the manure. In practice, higher dry matter levels are desirable so that the heap, seldom stored under a roof, when doused with heavy winter rain will not increase in moisture content so much as to become an unstable mass and begin to slump. For this reason SF should be stacked as high as possible to collect the minimum rainfall.

When stacked and left undisturbed, SF will spontaneously compost reaching temperatures in the range 30–50°C; the drier the fibre, the quicker does the temperature rise and to a higher value as shown in fig. 18, taken from further work by Pain *et al*. This data refers to heaps of SF only one metre high allowed to compost and temperatures measured. Large farm heaps probably attain higher temperatures and two days after commencement, a heap may be too hot for the hand to bear.

Once composted, SF remains inert and creates no real smell problem so that the material can be held in store without difficulty until the most suitable time for application to crops. This is undoubtedly an advantage, but more important is the lack of smell when spread on land—in contrast to whole slurry.

SF is, of course, a largely organic material and a few farmers have developed an outlet for it to the horticultural trade as a garden soil conditioner. Because of the variable nature of slurry and hence of SF, the tactic is not to sell as a fertiliser since minimum

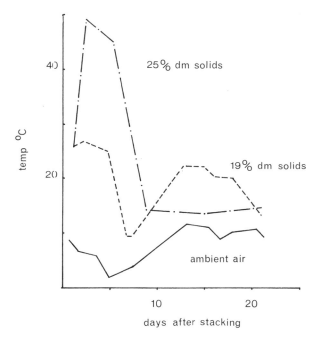

Fig. 18. Temperature of separated solids when composted.

average fertiliser content must then be declared and maintained, which leads to difficulties in operating a marketing outlet. For such sale, the composting should be thoroughly completed by careful mixing and turning; turned three times by a tractor loader seems to suffice for large heaps. It is also an advantage to have the SF as dry as possible because of its quicker composting and freedom from risk of any drainage whilst in the marketing chain.

On this question of the dry matter of the SF produced by a separator, sellers of machines can be induced to quite fierce argument over the value of the dry matter level. There are two views: that if one goes to the trouble of installing a separator, then the thing to do is to produce SF with as high a dry matter as possible (because it can withstand rainfall, will not seep and composts quickly); alternatively, there is no point in removing any more liquid from the fibre than is necessary for the fibre to be stackable and 'handleable' as a solid, indeed to remove more liquid simply means needing a larger liquid store.

The strength with which these views are put forward appears to be not unconnected with the type of separator being sold by the salesman. All machines produce a drier fibre when fed with a drier slurry and vice versa. Some machines are inherently unable to remove as much liquid as others, as illustrated in fig. 19.

Separated liquor (SL)

This looks like brown gravy and flows like water. Its dry matter content is much less affected by the DM of the original slurry and by separator design than is the case with SF (fig. 20). The ratio of original cow slurry DM to SL DM varies from 1·4–2·0:1, whereas slurry DM to SF DM varies from 1:1·4–1·5.

The different rheological (hydraulic) quantities of SL and whole slurry lie in the changed particle size distribution after separation. Whilst whole slurry may have 35 to 50 per cent of its weight made up of particles larger than 0·1 mm, SL may have only 5 to 15 per

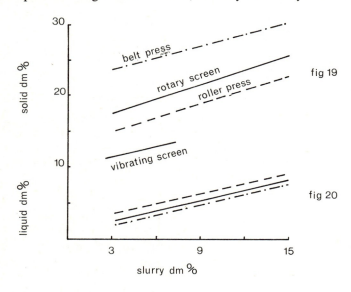

Fig. 19. Input slurry DM percentage and separated fibre DM percentage from various machines.

Fig. 20. DM percentage of resulting liquid from same slurry.

cent by weight of that size almost none of which would exceed 1 mm.

The free pumping nature of SL is maintained even after 18 months' storage in large containers. Cow SL may develop a very thin skin or crust on the surface, but pig SL does not seem to do this. Stored in bulk, SL is reported not to generate nearly the same level of odour as whole slurry and it is ideal for application to land through irrigation equipment whether fixed or mobile. However, some caution is needed here since the minimum rates of application possible with such equipment may be too high when the fertiliser content of SL is considered along with the needs of the crop to which it is applied. To meet this need for very even application of lower volumes of SL, a specially designed travelling spray bar irrigator has been marketed (Plate 27).

Whilst SL can also be applied by conventional slurry tanker, because of its free running character, it is possible to apply SL using a dribble-bar. An experimental attachment for a tanker was made at NIRD (Plate 29). This allowed SL to be applied evenly from very close to the soil surface, with a complete absence of aerosol generation and thus very little smell. Again, because of its nature, SL is rapidly absorbed by the soil surface, washes easily off plants and does not clog or shade foliage. This further reduces any tendency to persistent smell from a field following application.

To the livestock farmer, chasing tight stocking rates, there is often little grass available in late spring and summer for slurry application. One of the headaches is that of taint—discouraging cows from grazing—which can be noticed a month or more afterwards. Where SL has been applied, animals are reported to have voluntarily returned to grazing within a few days.

Whilst too early grazing after application raises the possibility of some risk from disease carry-over, SL runs into the bottom of the herbage and so gross contamination of the grazed foliage is unlikely. That animals will voluntarily graze is a sure sign that very little, if any, of the liquor remains on the leaves. It is known from comparative studies that application of whole slurry can result in changed behaviour of grazing cows who will drink more, lie down very much less, graze for shorter periods and eat up to 30 per cent less than where artificial fertiliser had been applied at similar rates of N. Hence the partial rejection by cows of pasture spread with whole slurry is largely avoided when SL is used and this must ease pasture management problems accepting that care over pathogen risk is necessary.

FERTILISER VALUE OF SEPARATED SLURRY

Work done at NIAE and NIRD on separation has produced information on fertiliser value. The mean values of several samples taken from cow slurries is shown in Table 20 and probably can be taken as a somewhat conservative 'average' level. Remembering earlier comments about the variability of manures and slurries, it will be no surprise that individual sample results from other sources show nutrient values sometimes almost twice as great.

Table 21 not only shows the amounts of SL and SF produced by separation of cow slurry, but also the distribution of plant nutrients resulting. The table contains results for thick and thin slurries and nicely illustrates that the DM of the original slurry greatly determines the nutrient content of the two fractions. It is thus evident how important it is to sample accurately a source of manure for nutrient analysis *before* planning its use as a fertiliser. Practical experience indicates that it is possible to get a much more representative sample from a large bulk store of SL than is ever possible with whole unseparated slurry. This is a further important practical advantage of separation.

Both slurry and SL are deficient in N and slightly so for P_2O_5, to be used unaided by artificial fertiliser as the sole source of nutrients for grass. Because of its free flowing nature, it is feasible

TABLE 20. Fertiliser Value of Separated Cow Slurry

	N %	P %	K %
Original slurry	0·23	0·04	0·15
Separated liquid (SL)	0·23	0·04	0·14
Separated fibre (SF)	0·24	0·04	0·14

TABLE 21. Plant Nutrient Distribution in Separated Cow Slurry

	Quantity (kg)	DM %	Total plant nutrients (kg)		
			N	P_2O_5	K_2O
Original slurry	1,000	12	3·4	1·6	3·2
SL	700	7	2·4	1·2	2·3
SF	300	23	1·0	0·5	1·0
Original slurry	1,000	6	2·3	1·2	1·8
SL	810	4	1·8	0·9	1·6
SF	190	17	0·5	0·2	0·4

TABLE 22. The Effect of Slurry and Separated Liquid on Grass Yield of Dry Matter

	\multicolumn{5}{c}{Total slurry or SL applied (m^3/ha)}			
	0	25		50	
		Slurry	SL	Slurry	SL
Equivalent N kg/ha	0	110	75	220	150
Grass yield total for 2 cuts dry matter t/ha	3·1	3·9	5·0	5·2	6·5

to supplement SL with liquid N fertiliser to make a tanker-load much more nearly a 'balanced' fertiliser. Perhaps only thirty litres or so of liquid N fertiliser need be added to a 5 m^3 tanker to achieve better balance of nutrients, and by so doing, the extra task of applying the additional artificial N required is avoided—yet another small but tangible benefit of separation.

Experimental evidence suggests SL is more effective as a fertiliser on grass than whole slurry. Both materials were applied on grass after cutting on two occasions during the season; SL gave up to one-third more herbage than the slurry even though the slurry contained more total N, as illustrated in Table 22.

SEPARATOR MACHINES

There are a variety of designs available including, in no particular order:

Centrifuge

Industrial processes quite often use centrifuges for filtration and separation and some years ago a design was tried out in agriculture. Throughput was not tremendous and the machine failed to gain a market largely on account of its price, although there could well have been reservations by farmers about the machine's cost of maintenance under corrosive agricultural conditions. Furthermore, it was introduced in the earliest days and so may have had to sell not only itself, but also the basic concept of separation which is a burden falling on any new equipment of a pioneering nature. Very few of the original dozen or so machines now remain active, but renewed interest is being shown by at least one manufacturer.

Vibrating screen

In essence, a perforated metal screen is caused to vibrate. The size of the perforations is varied, by fitting different screens, to suit the slurry to be separated. These machines are simple, robust and use little horsepower—0·75 kW for a claimed throughput of 6 m^3/h. There are designs featuring rectangular screens, double screens (one above the other) and models with 800–1,200 mm diameter flat screens. Vibrator separators are generally unable to deal with cow slurries with DM greater than 6 per cent and even then a very wet solid results, which produces a deal of seepage.

Pig slurry is dealt with more satisfactorily with little blockage and provided the fibre storage area includes a return drainage system to lead seepage back to the slurry reception pit, a workable installation results.

Roller-press machine (Plate 24)

The commercially available machines result from development work done at NIAE, wherein a conventional brushed screen is used as a primary stage on the machine followed by a roller-pressing section. Wet fibre is brushed from the first stage onto a semi-circular screen over which pass two spring-loaded rubber press rollers interspersed with two brushes. These brush out from the machine the squeezed fibre. Screen sizes used are usually 1·5 mm for pig slurry and 3·0 mm for cow or mixed slurries.

Throughputs vary with the slurry dry matter, but 1·5 m^3/h on cow slurry and up to ten times as much pig slurry are typical. SF DM lies typically in the range 15–20 per cent, rarely above.

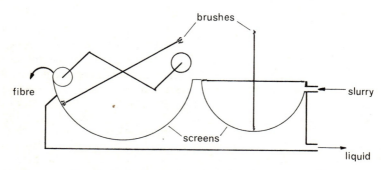

Fig. 21. Diagram of roller press slurry separator.

Belt press
In principle, slurry is distributed as a ribbon or swath on top of a mesh belt which travels between two press rollers—just like a clothes mangle, upon which in fact the first prototype machine built by Mr G. Shattock was based.

The woven mesh belt of man-made fibres is vulnerable to sharp objects such as chicken grit but the machine has an effective stone-trap to remove that hazard. Usually belt life is good, examples lasting over twelve months are not unusual.

The belt press is capable of producing a high dry matter fibre of 25–30 per cent DM which composts very readily. The theory advanced for this is that the metal screens of other machines are relatively much thicker and liquid can be trapped in the hole with some fibres as the roller passes, after which the liquid can be re-absorbed by the fibre. This cannot happen with the very thin belt and so more liquid is expressed and removed from the fibre.

Combined gravity screen and compression machines (Plate 26)
This is a two-stage machine in which incoming slurry is passed down a shaped wedge-wire screen through which much of the liquid is removed from the manure. The resulting damp sludge falls into a piston compression section where the hydraulically-driven piston forces the sludge into a cylinder having wedge wire walls. Escape for the sludge is out of the 'porous' cylinder past a stainless steel cone, the adjustment of which controls the escape area and hence the amount of pressure applied on the fibre by the piston.

With a slurry of 8 per cent DM, throughput will be under 2 m^3/h producing about 200 kg/h of fibre and the power consumption 5·5 kW. SF DM will usually be in the range 35–40 per cent, which is the driest fibre produced by any commercially available machine. Current price (1979) is about £7,500 exclusive of installation charges and provisions.

Siting the separator

Very often the site for a separator is dictated by the existence of a reception pit but where a free choice is available, points to bear in mind include:
- Separate as close to the animals as possible; this may save scraping time and unnecessarily large areas of dirty concrete.
- SL storage can be remote from the separator, SL can be pumped

long distances and this may allow the SL store to be positioned more conveniently for access during spreading time.
- SF storage can be either alongside the separation site or elsewhere, in which case direct loading of fibre into a trailer is necessary. The trailer must then be emptied each day or so. From a general farm management viewpoint, it is better not to engage in such a task daily, and so a more convenient arrangement is to let the SF pile up by the separator and use the tractor loader, say, weekly to move the SF into its long-term store nearby. As a guide, there will be about 1·5 m^3 of SF for each cow housed 180 days and unless a high reach loader or an elevator is available, a practical height of stack is 2 m or just over.

Feeding the Separator

This in practice is the biggest headache of all. The problem is to obtain an even supply of slurry to the machine despite the chance inclusion of a variety of materials in the manure and which might block or even break the pump. Clearly, the separator must be able to operate for long periods unattended since it is not acceptable to tie up a valuable worker just to watch for blockages.

Dilution

There is a saying that a separator will deal with anything a pump can supply. Pumps can operate with homogeneous semi-solids up to 14 per cent, in fact the limit very often is that the material in the pit will not run to the pump. However, practicality demands that the unseparated slurry is of low enough DM to be handled *reliably* by the pump system.

Pig slurry is often 6–7 per cent DM due to unavoidable water spillage from drinkers. It is worth noting that where cows are housed and fed under roofed accommodation, the scraped manure may be too stiff and some dilution will be needed to assist pumping. To add tap water costs money, and so the parlour washwater might be directed to the reception pit. An alternative is to arrange a return flow of SL from the main store as in this way the extra storage volume to cope with added tap water is not required. Best results are obtained if the dilutant is run into the pit before the freshly scraped slurry arrives.

The usual arrangement of underground reception pit with the

PLATE 24
A typical slurry separator installation. The separator is on top of the gantry so allowing gravity flow of liquor to the alongside storage tank and fibre to the heap on the right. Unusually, this separator is fed by a chain flight elevator.
                                                                                         Farrow Irrigation

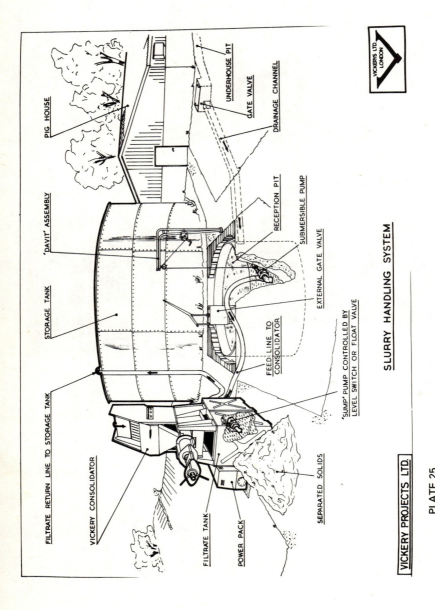

SLURRY HANDLING SYSTEM

PLATE 25
A diagram showing all the main elements in a slurry handling and separator installation.

Vickery Ltd

PLATE 26
This separator produces the driest fibre (40 per cent DM) of all. The primary gravity separation screen can be seen on top of the machine and the hydraulically-powered piston compression second stage.   Vickery Ltd

PLATE 27
A specially designed low-volume travelling irrigator for separated slurry liquid. The irrigator uses a turbine, driven by the slurry supply, to pull itself along a wire attached to a land anchor. The boom carries jets to distribute the liquid.   Author

PLATE 28
The same machine as in Plate 27, showing the width of boom and evenness of spreading liquid.
J. D. Pett and Sons Ltd

PLATE 29
A simply made dribble bar applicator for separated liquid. Evenness of application and near absence of smell generation are two outstanding advantages, yet no commercially made version is available.
Dr B. F. Pain, NIRD

separator mounted on a gantry is shown in Plate 24. Blockages are least likely if all the slurry is first of all chopped and homogenised in the pit. As mentioned earlier, to be able to chop large lumps of straw, etc demands a large orifice into the chopper pump, which in turn means a high power pump (15 kW is common) capable of large throughput (5·5 m^3/min), yet the separator can deal with only a tiny fraction of such a volume. It is clearly wasteful to run such a large pump for so long as the separator runs, so very often two pumps are fitted, the smaller to supply the separator but capable of providing a surplus flow to keep the pit contents agitated.

At least one manufacturer now offers an elevator (Plate 24) to lift slurry from the reception pit to the separator. The elevator body is a pair of box section plastic tubes 120 mm square within which the endless malleable iron link chain runs carrying reinforced rubber flights. Available in lengths 5 to 8 metres, speed is variable to provide up to 5 l/s volume to the separator and power requirement is only 0·75 kW. Where cows are bedded on sawdust or chopped straw, the elevator can probably manage unaided, but otherwise a chopper pump is needed to condition any long materials as the elevator cannot chop through lumps which may otherwise block it.

There is a common tendency for operators to run the separator at reduced throughput so as to avoid blockages. In fact, the separator itself can usually deal with its full rated output hour after hour and the operator's real concern is to prevent the separator slurry supply blocking. This normally occurs at the pump intake, and a common practice is to lead the slurry overflow from the separator back to a point very close to the pump intake to promote some hydraulic agitation at this point—every little helps.

To conclude this business of separation, the pros and cons may be listed as:

Disadvantages of Separation

- Extra capital cost.
- Extra running costs (power, repairs and supervision).
- Generally a regular daily operation.

Advantages of Separation

Separated fibre (SF).
- Handles like FYM.

- Composts readily.
- Very little smell.

Separated liquor (SL).
- Free-flowing and can be pumped needing only small bore pipes.
- Reduced liquid storage volume required.
- Liquid store can be remote from livestock buildings.
- No crusting problems in store.
- Bulk store of homogenous composition and can be sampled reliably.
- Rate and time of application to land more flexible and so less risk of pollution.
- Crop response to N in SL more reliable.
- Markedly reduced taint on grazing grass.
- SL can be irrigated at low volume and low trajectory, creating less smell.
- Feasible to balance up SL with liquid fertiliser.
- Smell control using minimum levels of aeration is possible.
- SL is likely to be a much easier material for anaerobic digestion (but not yet proven).

REFERENCES

PAIN, B. F. *et al*, Factors affecting the Performance of Four Slurry Separating Machines. *Journal of Agricultural Engineering Research,* **23** (1978), 231–42.

Chapter 8
AEROBIC AND ANAEROBIC TREATMENT OF MANURE

BOTH OF these approaches to the treatment of manure reduce the BOD demand, thus controlling smell. Aerobic techniques utilise oxygen, and anaerobic methods exclude oxygen to achieve results. They will each yield energy as a by-product, from aerobic treatment as low-grade heat and from anaerobic digestion as biogas containing mainly methane.

AEROBIC TREATMENT

The main reason for interest in aerobic treatment nowadays is for smell control, as it is now accepted that it is impractical and uneconomic to attempt to treat farm manures with a view to discharging to a watercourse.

The fundamentals of aeration are that a population of oxygen-consuming micro-organisms is developed and they feed on the organic waste for growth. The products are CO_2, water, heat and dead micro-organisms or sludge, and commonly 0·5 to 0·8 kg of sludge may be produced for every kilogram of BOD removed. The number of active micro-organisms is determined by the concentration of biodegradeable waste, the amount of oxygen available and, very important, the effectiveness of mixing within the treatment vessel or digester.

The methods of supplying oxygen vary and include:
- Compressed air supplied to a submerged porous diffuser which emits fine bubbles. Good efficiency of oxygenation but diffusers clog in manures.
- Jet aerator: the waste is pumped from the digester through a nozzle directed back into the liquid. Air is entrained and the waste mixed—like squirting a hose into a bucket of water. Good oxygenation efficiency and all equipment out of contact with the slurry, so easing maintenance. There is no commercially avail-

able equipment, although NIAE have reported favourably on the concept.
- Mechanical surface aerators, floating or fixed, which draw the waste through an impeller and throw the liquid into the air; oxygen then diffuses into the liquid, and more is entrained when the liquid falls onto the surface layer again. These tend to mix only a fairly shallow layer of the tank and are perhaps better used on lagoons when sludge settlement is less of a problem.
- Paddle or cage-wheels can also aerate but are not used because of their poor ability to carry solids.
- Aerators, usually floating, known as sub-surface down-draught types (Plates 33 and 34). Air is drawn down by the action of the spinning impeller and mixed with the waste. This has important advantages: surface turbulence and evaporation are at a minimum and so microbial heat is conserved further aided by a foam blanket on the surface; foam depth is controlled as surplus foam slides down the inlet cone and is forced back into the liquid. Available in sizes from 2 to 20 kW, their use can produce temperatures up to 60°C in the tank. This type of floating aerator is not prone to icing up in cold weather, which may cause other floating aerators to sink.

As with any biological system, maintenance in a steady state of the best conditions possible for the micro-organisms is important. In practice this means that shock loadings should be avoided. The sudden arrival of a volume of cold raw slurry into a digester will upset temperature distribution and locally will wash out biomass, both of which will slow down the overall activity. Any digester should be fed little and often and this can require some very careful thought about plumbing and control arrangements.

In practice many aerobic systems will be based on the tank or lagoon itself when the main precaution is not to completely empty the tank. 300 mm depth left in the bottom will act as an innoculum for the next batch of slurry to be added. Usually, such tanks are well buffered against trace chemicals in the slurry, but it is a sensible precaution to avoid large doses of hypochlorite etc from the dairy when a digester is starting up.

Heat Recovery

Since an aerobic digester can be operated at mesophylic (up to 40°C) temperatures or up to 60–65°C suitable for thermophylic bacteria, it is possible to recover heat from the process. In theory

14·5 kilojoules (kJ) are available for every gram of oxygen consumed by the bacteria. Presuming that 1 kg of oxygen can be provided in the waste for every 1 kW of motor power on the aerator, this means that 4 kW of microbial heat are available. In practice the likely level is probably 3 kW per 1 kW of aerator rating.

Formerly, this was of no concern but a number of farmers have expressed interest in heat recovery. Work on practically sized plants goes on in Denmark and Germany. Getting heat out by a heat exchanger system is straightforward, but to be of much use the grade of heat needs to be raised. To use water at only 30 to 50°C for heating involves large radiator surfaces and so the need is to increase the temperature by using a heat pump. This accepts a bulk of fluid at moderate temperature and produces a smaller flow of much higher temperature. At present there are no moderate capacity heat pumps available in UK at a reasonable price, but the idea can be made to work.

Aerobic Treatment of Solid Manure

Composting is a well-known horticultural operation and is a process whereby a loose textured heap of manure, usually mixed with straw, is kept damp and aerobic digestion takes place. The material is slowly converted to a friable, damp, stable non-smelling organic mass.

The idea was developed at Birmingham University by Drs Biddlestone and Gray to deal with pig slurry. A bin with a perforated floor was filled by hand with a thin layer of straw and slurry pumped on top. Liquid drained through the floor, was collected and pumped back over the straw. A small fan forced air at about 1 m per minute velocity up through the floor and the straw/slurry mixture. Rapid composting took place and a temperature of 60°C was attained. Loading was carried out each day until the bin was full, when it was left for three weeks to complete the composting action. By using four bins, a continuous process was possible.

The results were encouraging and so a compost unit was designed to serve a 1,000-pig herd in Staffordshire and another, of different design, to serve 500 pigs in Yorkshire. At present it is too early to finally judge the system. Like all new developments, there have been teething problems. Surprisingly these have been over simple things like how to evenly and automatically apply the slurry; how to ensure even loading of the straw without excessive handwork; and how to make a floor which is cheap, robust enough to withstand the high contact loads from a tractor and front loader during unloading yet is pervious to 'gravy' travelling downwards and air going upwards. Potentially the system could be attractive, in that liquid slurry is converted into a stable, non-smelling compost and no liquid left for disposal because of the compost warmth evaporating the water. For the pig farmer with little land and bothered about slurry disposal, the idea could be a lifeline.

Straw consumption is about 0·2 t/pig place/year and results in about 0·65 t/pig place of compost. Two hundred tonnes of straw for 1,000 pigs is quite high but it might be possible to work a straw for compost exchange.

ANAEROBIC TREATMENT OF MANURE

Since the first of our now recurrent oil crises in 1973, there has been tremendous interest by farmers and others in anaerobic digestion (AD) of farm wastes. Until recently, this interest was mostly on account of biogas produced, which is often looked upon as 'free' energy. The incidental smell reduction achieved by AD is now receiving much more attention, and whilst few farmers enquire on that basis alone, taken together with potential energy production, the combination certainly is more attractive.

AD as a subject merits a book in its own right and reference is made to some useful further reading. There is not space to consider the designing and building of a digester here, and so treatment is limited to some basic facts about the process and a discussion of its suitability for uptake by agriculture at the present time. Much of the information on AD in the UK has evolved from the work done at the Rowett Research Institute by Dr P. N. Hobson and his co-workers, who continue to research into this important topic.

Anaerobic Digestion

Animal excreta is a complex mixture of undigested food, gut bacteria and other intestinal secretions. To this are added other bacteria as the faeces travel to the collection point and ultimately into the AD digester. Some bacteria are aerobes and others facultative, *i.e.*, they can adapt to the presence or absence of oxygen, and some few are anaerobes.

In the closed digester, free oxygen in the atmosphere (of a newly loaded vessel) and in the water is used up by the aerobes (who then die or greatly reduce) and by the facultative bacteria. They and the anaerobes can then go on to digest anaerobically. The whole process is a dynamically mixed one comprising three stages: polysaccharides in the waste are hydrolysed to simple sugars; these are then converted into hydrogen gas and CO_2 to give some methane plus some acetic acid; and methane bacteria convert acetic acid to more methane.

Mention is sometimes made of the C:N ratio being important. Faecal waste contains more N than there is energy available for the bacteria to utilise all the N. This is academic under farm conditions but explains why large amounts of extra gas can be obtained by adding lawn mowings or potatoes to a working digester.

pH level

If for any reason, there are insufficient methane bacteria to convert the acid, the digester pH will drop from its usual pH 7–7·5 towards pH 6 as the acid content increases, eventually resulting in cessation of gas production as the methane bacteria are killed off. This is known as 'souring' and is one of the reasons why loading of a digester must be little and often. Where loading rate needs to be increased to cope with more animals, the extra load must be added in small increments over many days. For the same reason, starting up a new digester must be done with mainly water and a little waste which is gradually increased over maybe 8–10 weeks. In practice the new digester is usually inoculated with a few loads of activated sludge from a sewage works or other digester.

Temperature

Rate of digestion is better at higher temperatures, but in UK conditions 30–35°C is a good compromise between rate, stability and heating needs. It is much more important to hold the temperature steady within one or two degrees of the target level. In cool climates, much of the energy produced is needed to keep the digester warm and to raise the temperature of the incoming cold slurry. This may account for three-quarters of output in winter.

Heat is provided by heat exchangers inside or outside of the digester from a separate boiler fuelled by the biogas. Temperature lift across the heat exchange surfaces should be restricted to a very few degrees for the same reason as above. This means large surface heat exchangers.

Solids retention time (SRT)

The longer the waste is digested, the more complete the digestion but the cost of the vessel limits SRT. Ninety per cent digestion may occur in 20 days, but a further 20 weeks may not achieve 100 per cent. The minimum SRT is determined by the doubling time of the bacteria, *i.e.*, the time taken for bacteria to reproduce themselves. Too short SRT dilutes or 'washes out' the useful bacteria until digestion ceases. For AD three days is a minimum.

Gas quality

Biogas is a mixture of mainly methane and carbon dioxide plus traces of hydrogen sulphide. The proportion of methane varies with the animal species as shown in Table 23.

Digester

This may be above or below ground, airtight and insulated to reduce heat loss. It must be fitted with heat exchangers to keep it warm, a safety valve against excess pressure and a flame trap to prevent ignited gas burning back into the digester. There must be a stirrer (usually mechanical) which operates 5 minutes in the hour, a means of removing sludge and grit from the bottom and gas from the top. The slurry is pumped into the digester at the top by an external pump at frequent intervals (twenty minutes).

Apart from temperature control, the greatest problem is to prevent excessive crusting of fibre developing on the top. This can result in blocking off of gas pipes and even prevent gas from

TABLE 23. Anaerobic Digestion of Animal Manures: Biogas Production

Species	DM %*	SRT (days)	Reduction of BOD %	Production (m^3/kgTS)	BIOGAS Quality methane %	Volume per animal (m^3/day)	Gross* energy (kW/day)
1 cow	8	20	70	0·215	55–60	1·15	8·0
1 pig	8	10	83	0·300	70	0·12	0·8
1 hen	4	20	84	0·380	70	0·01	0·08

* DM % is the likely maximum suitable for use in a digester, for pumping or other reasons.

** Gross energy is theoretical maximum from which should be deducted the variable amounts of energy needed for warming the digester vessel and the cold untreated manure.

escaping from the liquid in the digester. The usual arrangement is to control liquid level by simple overflow through a gas trap weir box which will also set the pressure at which the gas system will operate (100–200 mm of water pressure). Very often the gas supply may be fed to a small gasometer of 5 or 6 m^3 to stabilise gas flow and pressure.

Sizing a digester is determined by SRT and the number of animals served. Details of salient factors of performance for the three main species are shown in Table 23. For example, a digester vessel for 1,000 pig places would be sized as follows:

Pigs excrete manure at 10 per cent dry matter but it is common to find pig slurry at 6 or 7 per cent. A fattening pig produces 4 l daily of excreta which at 6 per cent DM would amount to 6·4 l

1,000 pigs would produce 6,400 l.
SRT of 10 days for pigs (Table 23) ∴ volume required 10 × 6,400 l.
Allow extra 20 per cent on digester volume to provide gas collection and short-term storage: $10 \times 6,400 \times \dfrac{120}{100} = 76,800$ l or $\dfrac{76,800}{1,000}$ m^3 = 76·8 m^3 (4 m diameter × 6 m high).

This is quite a small digester but it would produce over 20 kW hourly. Scottish work on pig slurry suggests a rule of thumb: 0·45 kW/m^3 of daily slurry throughput at 6 per cent dry matter.

The above is an arithmetic example since in practice sizing is more complex.

Should the digester be capable of treating all manure produced including the winter maximum? If not, what does the farmer do with the untreated manure surplus which may arise from time to time? Will he enjoy capitalising and managing two systems of manure handling? Is there likely to be any fluctuation in animal numbers? Might there be herd expansion? Should extra emergency storage for untreated manure be provided to cater for fluctuating supply of slurry?

Should the digester be oversized somewhat to be able to recover any lost throughput and, if so, by how much? Simple practical questions, which are often overlooked yet need consideration before the final size of digester is settled.

UTILISATION OF BIOGAS ENERGY

To utilise this energy to save money on the farm is the acid test and much more difficult to accomplish than many believe. There are two options:

(a) burn the gas in a boiler to produce hot water for use about the farm or to heat the farmhouse;

(b) use the biogas to run a spark ignition engine driving an electrical generator.

(a) Biogas can be burned directly in a hot water boiler. The burner needs to be the type supplied for town gas *i.e.*, running at 100–150 mm water gauge pressure and having large orifice-holes. A grey or yellow/grey deposit from biogas develops in the burner but is easy to brush out.

The problem with hot water production is that it is expensive to connect to more than one or two consuming points. In summer, when most gas is available, there is least demand and much more of the gas cannot be put to use.

(b) Much more flexible in its potential transmission and use is electricity, and this attracts most enquirers especially with electricity bills on some farms now running into five figures annually. The gas can power an engine-driven generator and the current used anywhere on the farm. There is no likelihood, however, of electricity authorities contemplating surplus power being fed back into the public supply, indeed it is possible that increased standing charges may be levied on a farm which reduced its mains consumption. Modified stationary engines to run on methane are available and the waste heat from engine cooling water and exhaust is collected and fed to the digester, normally being more than enough to keep up temperature.

Recently, the Fiat Company released onto the UK market their TOTEM unit (total energy module) which comprises an automotive 903 cc engine modified to run on biogas with built-in heat recovery from cooling water, block and exhaust. An electrical generator can be built in and the unit can then provide power and hot water. The overall recovery of energy from the fuel (biogas) consumed is over 85 per cent.

Having produced power, there are problems. Will the farm power always be available and, if not, can mains back-up be automatically provided? Which of the many farm circuits should be connected to the generator? How big can the generator be—almost certainly gas supply will be limiting—and how reliable will be the gas supply? It will be necessary to know just how much the extra costs for control gear and its maintenance will be, and there will be a need to supervise the working. These few simple questions begin to illustrate that the greater problem is not to produce gas but to use it! It is almost universally overlooked that *reliability* of mains electrical supply is always taken for granted, but to achieve the same standard with home-produced power demands extra effort and organisation.

Safety

Methane forms an explosive mixture in the range of 5–15 per cent methane (7·5–22·5 per cent biogas) in air, so there is some potential safety risk.

At present there are no specific regulations covering the operation of a digester although the general provisions of the Health and Safety at Work Act, through the Agricultural Regulations, will apply. Those contemplating installing a digester ought to discuss any requirements with the local Agricultural Inspector. The farm's insurance assessor will also need to know.

Digesters run at a positive pressure, so an in-leakage of air is not possible. Air could be pumped in by a pipe leak on the suction side of the slurry dosing pump, but only small amounts are likely which would be used up by the facultative bacteria. A large leak at the pump should be noticed during the daily inspection.

Leaks below the liquid level in the digester will be of slurry not gas. Leaks of gas from above the liquid level will be at high level in above-ground digesters, and as methane is lighter than air, leaks will diffuse upwards.

Perhaps the greatest risk is the gas supply pipe from digester to point of use. Careful planning and high standards of installation and protection are necessary. In addition, a flame trap is always fitted on the supply pipe at the digester. Damage by accidental vehicle impact could rip open an above-ground digester, and so siting of the digester merits consideration and perhaps strategically-sited bollards or crash fenders.

Plant Operation

Ease of running the plant must be a high priority, as farmers are busy and involved in many other technologies. Few farms will have enough labour to allow someone to specialize to any extent on the running of an AD plant. Due to circumstances on the farm, many farmers may not make much in the way of concessions to an AD plant. It will be expected to run with little attention and to be very reliable month in, year out.

It is sobering to reflect that under farm conditions, animal excreta is but one element of a material which may be collectively described as 'raw effluent'. The likelihood of stones, wood, plastic sheet, silage, bedding, straw, metal and so on must be anticipated by the designer. This is a tall order if an AD plant is to be regarded as even reasonably trouble-free in practice. Thus considerable design ingenuity is necessary.

The plant should be self-monitoring in all aspects, self-correcting and, above all, fail-safe with the ability to decide when to call for help from the farm staff. It is perhaps just too early at present for the new micro-processor technology to be of direct impact here, but clearly this area of truly automatic unmanned monitoring must be a high priority for future design.

There seems to be a strong case for the buyer of an AD installation to also be able to obtain a contract service/maintenance/fault rectification back-up arrangement. This may be difficult and expensive to provide, at least until AD plants are more numerous.

Costs

This is *the* question and very difficult to answer at present as there is a lack of performance information for full-scale digesters operated under practical farming conditions. Currently there are fewer than a dozen farm plants in the UK, and less than half are moderate to large sized digesters which have not yet run long enough to provide useful information on overall performance.

However, as a discussion starter, consider a 5,000 pig-place farm which decides on a digester and electricity production. Digester size: 5,000 animals @ 4 l/d = 20 m^3 at 10 per cent DM or 33 m^3 at 6 per cent DM for pumping, plumbing and other reasons. 10 day SRT and 20 per cent extra digester volume for gas and crust means a vessel of 399 m^3.

On the basis of 0·3 m^3 biogas/kg total solids added and a calorific value of 6·7 kW/m^3, daily power produced would amount to 4,020 kW of gas energy. Converted to electricity at efficiencies of 0·3 for the engine and 0·75 for the generator, suggests a continuous output close to 38 kW/hour. Assuming mains power costs 3p/kWh, the home-produced electricity is potentially worth £27 daily *if* it could all be used to save buying mains power.

'Turnkey costs' of such a digester, including the generator set but excluding any slurry or digester effluent storage capacity, are likely to be £45,000. Of the installation, some elements will be very long lived, *e.g.*, the tank, and can probably be written off over, say, 20 years. There will be other elements like pumps and other machinery which will need renewal at intervals, certainly within 10 years. Thus for investment purposes, of the £45,000, £20,000 might be written off over 10 years and the remainder over 20 years.

General maintenance and repair of the plant (except the engine/generator unit) may be set at 3 per cent of capital: £1,350.

Estimating the generator to run 350 days of the year continuously would amount to 8,400 hours' use, and so the power unit would require an annual complete overhaul. The cost of this is, of course, conjecture because engine life under constant steady load, with scrubbed gas fuel should be very good—bottled-gas-powered fork-lift trucks demonstrate this. However, as a guide, an 80 kW tractor engine after 5,000 hours would require a major overhaul, for which a figure of £1,000 could be anticipated, to which should be added transport and removal/refitting charges. Including a sum for generator repair, then perhaps £1,200 might be considered a minimum cost. Reliability being so important, a new engine alternate years may be necessary, in which case the annual cost would increase to £1,600.

Some cost must be set against supervision which may occupy a man for say 7 hours weekly, 50 weeks of the year, and he would assist in the annual overhaul for 2 weeks. This would total 438 hours a year costing £644 (tractor driver level of £64·73p for 44 hours). In addition, there would be a need for extra payments for 'call out' duties—say, 6 incidents per year averaging 2 hours = 12 hours at time and a half rate (£26). Total labour cost £670.

No costs for electrical control gear included

No costs have been included for the purchase and installation of electrical control gear to connect circuits about the farm buildings. Similarly, no costs are listed for any necessary diagnostic consultancy which may be needed to deal with digester problems.

Finally, there may well be a cost set against an AD electricity generation plant if the farms connected load to the mains supply is reduced. It is likely that supply authorities would seek an increased quarterly standing charge for such farms.

Thus, a table of the costs can be drawn up as follows:

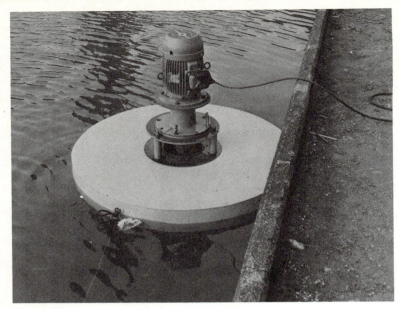

PLATE 30
A floating surface irrigator designed to throw liquid into the air for aeration purposes.
Farrow Irrigation

PLATE 31
The aerator from Plate 30 in action.
Farrow Irrigation

PLATE 32
A large surface aerator at work on calf manure. It is easy to see how in frosty weather, this type of aerator may ice-up sufficiently to sink unless switched off during cold weather. *Author*

PLATE 33
A floating sub-surface downward draught aerator. Air, and foam, is sucked down the cone and mixed into the liquid by the impeller. A spare impeller can be seen propped against one of the legs. This type of aerator controls foam depth to a constant level. *Author*

PLATE 34
The 14 kW aerator from Plate 33 at work in pig slurry. In an uninsulated chalk lagoon, liquid temperatures over 35°C were maintained comfortably throughout the cold 1978/79 winter. The warm vapour clouds are clearly visible and attempts by the farmer are under way to use energy from the lagoon to heat the piggery.
Author

PLATE 35
A 350 m^3 anaerobic digester treating pig slurry from 5,000 pigs. Biogas is used to power a spark ignition engine generator set housed in the acoustic cabin on the right.
Farm Gas Ltd

PLATE 36
The inside of the generator hut (Plate 35) showing control gear and the 35 kW 3 phase generator. Farm Gas Ltd

PLATE 37
A below-ground anaerobic digester treating manure from a 320-cow herd. Of unique design (by Helix Multiprofessional Services), the insulated glass reinforced plastic sectional floating roof acts as the gas store too. The biogas is used to generate 3-phase electricity. Author

Capital Cost of 400 m³ Digester including Electric Generation: £45,000

	9 % interest	18 % interest
Annual cash flow of write-off (ignoring the marginal tax level of the farm)	6,189	9,482
Maintenance and repair general plant	1,350	1,350
Maintenance and repair engine/gen	1,600	1,600
Supervision	670	670
Balancing of electric circuits; automatic control gear; installation & commissioning	Variable	Variable
Additional services required (diagnosis of problems)	Variable	Variable
Increased standing charge for continued mains connection	Variable	Variable
	£9,809	£13,102
Power generated, theoretical maximum & assuming total reliability & 100% utilisation of all gas produced		£9,497

Perhaps these figures will provide food for discussion, not to say even argument. Their basis is as quoted and several sources of information and experience have been sounded for cost and performance levels. At a cursory glance, the table suggests that financially the digester could be in the area of breaking even, much depending on the interest rate incurred by the investment.

It really is quite vital to stress that these figures are not based on actual performance levels. The more practically minded reader will recognise the enormous presumptions included in the calculations:

1. That reliability will be absolute—the lesson from the water and electricity industry is that where sustained power loads are demanded of engines with a high degree of reliability as well, slow running heavily constructed engines are necessary. These are very expensive and the costs prohibitive in farming, therefore the use of high speed engines for farm generation plants suggests frequent renewal.
2. That yield of gas will be 100 per cent of the research findings.

3. That utilisation of the gas will be 100 per cent throughout the year—almost certainly achieved utilisation will be some way short of this, but what level can be expected—25 per cent, 50 per cent, 70 per cent?
4. That farm staff can manage the plant unaided throughout the year.

To emphasise the importance of performance and utilisation levels achieved, if gas production were 80 per cent of indicated laboratory levels and only 50 per cent of gas could be usefully put to work, then the value of power produced would drop to £3,799. This would leave a deficit of £9,303 on the year's operation.

It would be easy to write-off biogas production as a result, but great caution is needed over the figures because as stated before:
- Performance levels are theoretical.
- Capital cost is an estimate based on early 1979 prices inflated by 17 per cent.
- There would be an element of tax reliex depending on the individual farm circumstances.
- The economic situation is dynamic; world energy prices are moving upwards which is an increasing asset to a digester operation.
- the degree of smell control is a potential real asset to a pig farmer but financially unquantifiable in the general case.

All these factors influence the judgment about the financial viability of anaerobic digestion and so much is dependent upon the time scale over which the investment is to be viewed. It is a matter of conjecture as to the cost, in real terms, of the mains electricity in 10 years' time, which would be approximately only half the working life of a unit built now.

REFERENCES

HAWKES, D. L., HORTON, H. R. and STAFFORD, D. A. Methane Production from Waste Organic Matter. C. R. C. Press, U.S.A. (1980).

HOBSON, P. N., et al, Anaerobic Digestion of Organic Matter. *C.R.C. Critical Reviews in Environmental Control,* **4** (1974) pp. 131–91.

HOBSON, P. N. and ROBERTSON, A. M. Waste Treatment in Agriculture. Applied Science Publishers.

Chapter 9

UTILISATION OF MANURES AND WASTES

THE POTENTIAL for the re-use of manures and wastes is theoretically quite large, but space prevents a complete discussion here. The scientist can point to the possibilities of producing protein from wastes following aerobic treatment with a dash or two of chemical cocktail, then letting it mature with some sunshine in a lagoon, harvesting the algae and drying it to produce a protein meal. Another favourite theory is to grow water hyacinths on waste lagoons as a means of producing biomass material.

No matter how ingenious the idea, so many of the recycling theories fall down over the hurdle of economics, *i.e.*, the end product costs more than other more readily available sources. It is right that scientists should be interested in new concepts, but the practising farmer as a businessman must obey the laws of economics or go bankrupt. Hence many ideas for recycling wastes are best left well alone until proven techniques are available and circumstances have changed to favour their economic costs.

Therefore, to confine comment to proven techniques of utilisation reduces the field to the use of dried poultry manure, ensiled poultry litter, aerobic and chemical treatment of pig manure and, of course, the age-old practice of using manure as a fertiliser.

MANURES AS FERTILISER

As Table 14 shows, the nutrient content of manures has a financial value as a substitute for bought-in artificial fertilisers. A few examples follow to illustrate how manures can be integrated in the fertiliser programme. The data is extracted, with permission, from one of the excellent series of Farm Waste Management booklets produced by the Farm Waste Unit of MAFF.

The examples should not be followed blindly since crop nutrient requirements vary with soil type, weather and soil nutrient status, which in part is determined by the previous cropping and fertiliser policy.

EXAMPLE 1—*FYM from cattle on main crop potatoes*

	N	P_2O_5	K_2O
Nutrients required kg/ha	220	300	300
Supplied by FYM (Table 13) 50t/ha	75	100	200
Additional art fertiliser req'd	145	200	100

With potatoes some excess of nutrients over the basic recommendations is tolerable.

EXAMPLE 2—*Poultry slurry on kale*

	N	P_2O_5	K_2O
Nutrients required	125	75	75
Supplied by slurry (Table 13) 14 m³/ha	127	77	76
Additional art fertiliser req'd		NIL	

Where all the N is to be supplied by the slurry, it is very important that the slurry quality is good enough and not reduced by long-term storage or by dilution. This is a situation calling for laboratory analysis of nutrient content *before* application so that application rate can be adjusted if necessary.

EXAMPLE 3—*Cow slurry on grass grazed by cows*

	N	P_2O_5	K_2O
Nutrients required for whole season	400	30	30
Supplied by slurry (Table 13) 7 m³/ha	18	7	32
Additional art fertiliser req'd	382	23	NIL

The immediate reaction here is: why not use more slurry? The application is limited by the potash situation, whereby if excessive potash is applied to grazing, the grass may become deficient in magnesium content leading to hypomagnesaemia in cows which graze the crop. This possible risk of upset can become a fetish sometimes and where circumstances suggest the herd is at risk, a magnesium-supplemented dairy cake can be fed.

From the example given above, not much slurry can be spread on intensively grazed grass and the problem on the heavily stocked cow farm of how and where to spread slurry becomes evident. Other complications are taint from slurry spread on grass, which may drastically reduce dry matter intake by the cow, and the need to guard against pathogens by not grazing too early after slurry spreading. For the dairy farmer, the better place to apply slurry is on silage or hay aftermaths and also after very close grazing. There is then sufficient time available for taint, etc to dissipate before the next grazing.

To complete comment on example 2, 100 kg N can be applied in spring followed by 75 kg/ha N after subsequent cuts or grazing during the season.

EXAMPLE 4—*Cow slurry on grass cut for silage*

	N	P_2O_5	K_2O
Art fertiliser only early spring	120	—	—
Art fertiliser for first cut	120	60	90
Slurry applied *immediately* after first cut 20 m³/ha	50	20	90
Additional art fertiliser at same time	50	40	—
Art fertiliser for second cut	100	30	60
Slurry applied *immediately* after second cut, 13 m³/ha	32	13	58

This would leave a debit of 17 kg/ha of P_2O_5 to be included in whatever the continued use of the field and the summer weather indicated for fertiliser use.

Similar arithmetic can be done to determine the amount of slurry which can be applied to spring cereals. Practically speaking, the N fraction should not all be supplied by the slurry lest the young crop runs short of N early in its life. It is prudent to use artificial N fertiliser for a quarter to a third of the total need. Another consideration is that a heavy tanker may leave ruts in the field which cannot always be removed. Such ruts can occasionally make life difficult for subsequent spraying and even combining operations.

MANURES FOR RE-FEEDING

Not all manures are suitable for re-feeding to animals. Ruminant animals eat roughages and tend to extract rather more of the nutrient content of their food than monogastric species like the hen and the pig. Furthermore, the economics of modern livestock farming dictate high energy rations for pigs and hens to encourage fastest growth rates. Poultry droppings contain compounds produced within the hen's digestive process, particularly nitrogenous components like uric acid, microbial organisms and, of course, wasted feed. Thus their manure is rich in nutrients and may contain a quarter or more of protein which can usefully be re-fed.

All re-feeding of faeces requires control of pathogens, parasites, etc and most of the research and development work has gone into this crucial aspect. At present there is a minimal amount of re-feeding of manures in UK, although interest has quickened in the last three years in tune with the energy crisis and its resulting generally wider interest in the recycling of resources. In contrast, in USA re-feeding of all types of manures has been practised for three decades on some farms. Much of the practice has been carried out on the large feed-lot operations where perhaps several thousand animals may be fed in yards in systems quite unlike UK conditions. American experience has shown that, given good management, quality carcases can result.

There are two benefits by re-feeding manure: some reduction of final volumes of manure produced by the livestock enterprise; and more profitable use of the manure.

The first statement should be treated carefully as, although true, it is bordering on the academic from a practical viewpoint. A 15,000-bird broiler shed producing, say, 300 tonnes of broiler litter (straw and droppings) annually would support 465 beef animals (fed a ration containing 30 per cent broiler litter) which in turn would excrete about 1,100 tonnes of manure or six times the original volume of broiler litter. This would result in a much bigger manure production on the particular holding if the beef enterprise were added to utilise the poultry waste.

The real incentive to recycle manure is financial. As a fertiliser, the greatest value poultry manure can attain is £4·53 per tonne (Table 14), whereas it could be worth well over £60 per tonne even when diluted with litter and fed to beef animals, saving 15 per cent of feed costs.

There are two methods of killing off pathogens in poultry manure before re-feeding: by drying or by acidifying, *i.e.*, lowering the pH sufficiently to achieve pasturisation.

Manure Drying

It is as well to distinguish between dried poultry droppings from battery birds, referred to as dried poultry waste (DPW) and dried poultry litter (DPL) where the birds have been kept on litter. This has a higher fibre content as the droppings are mixed with litter, although there may be a higher true protein content because of microbial action in the litter on the uric acid. Figures from the Poultry Research Centre at Edinburgh show an average dry matter for DPW of 92 per cent and 27 per cent crude protein. This can be compared with DPL at 85 per cent DM and 25 per cent crude protein.

Dried poultry manure (DPM) whether litter or waste, can be fed to cows, beef and sheep. It has been successfully included up to 30 per cent in compound feeds for ruminants, which can more effectively use the uric acid than pigs or poultry. The energy value of DPM is low—about that for hay or straw—but it does contain minerals, especially calcium and phosphorus.

For pigs and poultry DPM is of lower feed potential because they can use neither the uric acid nor the fibre. DPL is less useful than DPW for monogastric stock as they cannot deal with the fibrous litter. DPW is broadly equivalent to barley on a protein basis but only about one-third as good as barley on an energy basis.

Hazards

The re-feeding of poultry manure poses questions as to the possible presence of other materials which could be harmful. Litter, heavy metals and medicinal residues may be present as well as disease organisms. The high temperatures reached during drying effectively control pathogens provided the drier is operated correctly and not overloaded with wet manure. Particular care is needed to ensure that no cross contamination occurs between the pathogen-free freshly dried manure and the raw material awaiting drying.

Growth promoters and medical products are seldom used at high rates for laying hens and so DPW is unlikely to contain medicinal residues. DPL is better fed at lower rates of inclusion in a ration as it may contain some coccidiostats which survive drying; myco-toxins from wood preservatives present in the original wood from which sawdust or shavings litter is generated and perhaps insecticide residues may be present, too.

The foregoing sentences are included largely for interest's sake because manure drying has reached an all-time low nowadays. In 1972 there were about 100 driers in operation. The following year saw the first of our recurrent oil crises which doubled the cost of oil. At about the same time, the number of complaints about foul smell from driers had begun to increase. The smell problem was overcome by fitting the drier with an after-burner which increased oil consumption by 70 or 80 per cent. These two extra costs coming so close together and the increasing readiness of people to complain about smells has discouraged most farmers from drying manure. In 1979 there were thought to be about 24 plants in operation, only ten of which were in full-time use.

This means that it is rare that DPM will be offered for sale as feed since most dried material is produced by the larger poultry enterprises who are big enough to be able to use all the product themselves. They therefore do not have to meet the regulations governing sale under the Agriculture Act 1970, the Medicines Act or the various schedules of the Fertilisers and Feedingstuffs Regulations 1973. Certainly where a farmer were to contemplate either drying manure or feeding DPM he would be very well advised to seek veterinary advice before so doing.

The long-term outlook for high temperature drying is poor mainly because of the very high cost of fuel. Facing as we do, rising energy costs, any system demanding high energy inputs is difficult to support and impossibly so when there are alternative minimal energy techniques available as is the case here with re-feeding poultry manure.

Ensiling Poultry Litter (EPL)

Apart from drying, an alternative way of controlling pathogens, chiefly salmonella, is to increase the acidity of the litter until pH reaches 4·5 and to hold in this condition for six weeks usually by ensiling and covering. Formic and propionic acids have been tried successfully on a laboratory scale but the rate of addition (four per cent) was so high as to make the process rather expensive.

American work, confirmed in UK laboratory studies, has shown that a six-week period of storage of used litter under oxygen-free conditions effectively controls pathogens if the moisture content of the litter is adjusted to 50 per cent. Since the process depends upon anaerobic fermentation to produce acids which lower the pH sufficiently, the best chance of success will be where there is sufficient energy present to provide a substrate for the lactobacilli present in litter to feed on and multiply quickly to produce the acids required. Artificially inoculating the litter with bacteria was shown to have no advantage.

A number of additives have been tried such as water alone, molasses, ground barley and ground barley/malt mixtures.

Adding water only

This was frequently successful in achieving low PH levels but not always so. The amount of wasted feed in the litter seems to be the key here and obviously will vary from batch to batch. Generally, a better fermentation was achieved with wood-shavings-based litter than with straw-based material. A number of farms follow this practice without problems, although it is most important to obtain a good seal on the silo.

Adding molasses

This is a cheap additive which also enhances the nutritional value of the final product. Rates of inclusion from 5 to 20 per cent have been examined and about 10 per cent addition looks to be reasonable. However, molasses is very difficult to handle so as to be able to apply it intimately with the manure and moreover its price can fluctuate.

Adding ground barley and barley/malt mixtures

One part high diastase barley meal diluted with five parts ground barley meal when added at the rate of 50 parts of the mixture to 100 parts wood shavings litter at 50 per cent moisture content resulted in pH 5·0. However, by omitting the expensive malt and using only ground barley meal, satisfactory results have been achieved using 50 per cent inclusion on a dry-matter basis. The ground meal can be easily metered by auger into the litter at the point of ensiling.

Technique

Although the ensilage process controls pathogens, it is prudent to have samples of poultry litter examined for the presence of salmonella organisms beforehand and, if present, the batch should not be used.

The litter should be watered to reach 50 per cent moisture content. As found in the broiler shed, litter is often only 30 per cent damp and so it must be sampled to determine the amount of additional water necessary can be calculated. Very often about 25 mm of water will be needed which can be applied either by hose, overhead horticultural sprayline or tractor sprayer in the broiler shed. Alternatively, if the litter is to be handled on a conveyor or by blower, that may be the best opportunity to add water. The litter needs to be uniformly damp and so a period of time may be necessary for all moisture to be absorbed.

Ensiling methods

Tower silos are theoretically ideal, especially the top-loading/bottom-unloading variety so popular in the USA. However, a powerful blower is necessary, and propelling the damp litter to the top of tall silos is problematical as is achieving even and consistent loading in the tower. Much more work is necessary to develop this method under UK conditions and investigations continue to be carried out by at least one manufacturer of top-unloading tower.

The plastic sausage produced by the well-known Eberhardt Silopresse is quite convenient and lends itself to a contractor service. This machine, designed for grass silaging, packs chopped material into a plastic film (0·2 mm thick) envelope about 3 m wide and 1·75 m high. A 50-tonne load measures about 15 m long, having a calculated density of 700 kg/m^3. This would take 3·5 to 4 hours to produce.

Successful results have been achieved with this method, but there is a slight reservation in that the white plastic material is slightly pervious to oxygen and so the outer 100 mm of silage is not strictly in anaerobic conditions. It would be possible for pathogens to survive in this layer, so the merit is obvious of checking for salmonella *before* ensiling.

The simplest method of ensiling is to load the litter and ground barley into an airtight surface silo. On the evidence of investigations and farm experiences so far, best results are in a moderately deep silo, say 2 m. The sides and bottom should be covered with 500 g polythene, the litter loaded in by elevator, blower or tractor shovel depending upon the wetting and mixing system and the whole silo covered with a weighted top sheet which should be painstakingly sealed to the bottom sheet.

Whilst pathogen control is achieved in the clamp, it is very important not to nullify this by permitting cross-infection from litter left lying about the silo site. All dust, droppings and spilled litter should be carefully removed after the silo is sealed.

Feeding EPL

EPL has been fed to 3-month-old calves successfully and to beef animals from 150 kg liveweight at up to 50 per cent of the ration without ill effects. Beginners probably ought to stick at 15 or 20 per cent EPL in the ration but even at this rate economies in feeding costs of beef look to be around the 15 per cent level (£25 per animal) which is attractive.

There are no indications whatever that the meat produced is in any way detectably different from conventionally-fed beef. Indeed a Scottish farmer has built a demand for EPL-fed carcases and has a throughput of several hundred animals a year.

Anyone contemplating EPL production and feeding would be well-advised to seek competent nutritional and veterinary advice as there are precautions to be observed. The pathogen risk has been mentioned, and apart from ensuring the correct ration, the nutrition chemist can check the trace element and mineral contents too.

Your veterinary practitioner will explain any restrictive legislation which may affect the feeding of poultry manure in the future. Since 1972 there has been a proposed Protein Processing Order under discussion. Its aim is to prevent re-cycling of salmonella, etc in animal, bird or fish material processed for inclusion in livestock feeding stuffs by licensing such plants. For the time being, the order would not apply to the processing of material produced by, and intended for feeding to, livestock or poultry kept on the premises. However, it would apply to material transferred from one holding to another either under the same or different ownership.

This order is not yet in force (1980), but it would be sensible to check any likely implementation with the state veterinary service before investing finance in an EPL operation.

At present, veterinary opinion is against the feeding of EPL to dairy cows since if salmonella did survive ensilage, the pathogen could very easily get into the human food chain in the milk.

PIG MANURE

The re-feeding of pig manure is quite feasible given pathogen control. A deal of research has gone on in America, yet there has been little real interest in the topic by UK pig farmers. During the running of a long series of experimental observations on pig waste treatment at the MAFF Experimental Husbandry Farm, Trawscoed, in Wales, some of the resulting final clarified liquor was given to the pigs in lieu of drinking water. After a while, the pigs began to show signs of nitrate poisoning and so mains water was restored for drinking.

When slurry is agitated and supplied with air, a micro-organism

PLATE 38
An anaerobic digester built by a Dutch farmer. manure from 2,700 pigs is treated and the biogas used to produce hot water for heating purposes only. The (black) floating roof is of all welded construction and acts as the gas store. The digester tank is of rendered reinforced concrete construction. Digested effluent is stored in the lagoon.
Author

PLATE 39
The Fiat TOTEM unit: 903 cc car engine modified to run on biogas and connected to a 3-phase electricity generator. Waste heat from the engine exhaust and cooling water is recovered via heat exchangers as hot water. Overall energy efficiency of the unit is claimed to be 85 per cent.
Fiat Motor Co. (UK) Ltd

PLATE 40
The Fiat TOTEM unit showing the 3-phase 15 kW electricity generator and one of the waste heat exchangers.

Fiat Motor Co. (UK) Ltd

PLATE 41
A bank of TOTEM units running on biogas at an Italian sewage works. The concept is that as power demand fluctuates, so an automatic controller switches in or out one or more TOTEM units.

Fiat Motor Co. (UK) Ltd

PLATE 42
Drying poultry manure beneath laying cages. A fan recirculates house air via the polythene duct, positively over the manure to aid drying. Manure of 50 per cent DM results. Fan power is rated at 0·25–0·3 W/hen at 10 mm water back pressure using fans of 300 mm diameter running at 2,800 rpm.

Ing. W. Kroodsma, IMAG, Holland

PLATE 43
Another way of encouraging poultry manure beneath laying cages to dry out. The ceiling type fans force the warm ventilating air over the manure and so increase the evaporation of moisture. Fans spaced 8 m apart and providing 3 m³/sec proved adequate.

Ing. W. Kroodsma, IMAG, Holland

PLATE 44
A dribble bar attachment for a Farrow slurry tanker. Large diameter pipes are used so that whole slurry can be handled. The quick release clips to enable the distribution manifold to be cleaned out easily can be seen. To shut off slurry flow positively, each pipe is nipped and raised by the action of the triple bar frame shown held in its out-of-work position by the latch. Farrow Irrigation

PLATE 45
The Weeks slurry curtain attachment for tankers. Slurry is passed into the box section crossbar manifold and down between two plastic sheets to reach the ground. No splashing or aerosoling of slurry occurs and very little smell results. Weeks Trailers Ltd

population develops which consumes the organic matter and releases heat. If the vessel is insulated to reduce heat loss, the temperature rises and by operating at elevated temperatures the pathogens are killed off.

At the national demonstration, Muck '77, held at the National Agricultural Centre, Stoneleigh, the Department of Agricultural Engineering at the Universtiy of Newcastle upon Tyne revealed encouraging results from their investigations of aerobic digestion of pig slurry. Briefly, pig slurry of 4·5 per cent dry matter was aerated for two days at a temperature of 55°C which increased the amount of true protein (*i.e.*, nutritionally available to an animal), reducing the COD by 40 per cent *and* eliminated salmonella after the first four hours. Least cost computer studies of incorporating the resulting treated effluent into a pig fattening diet showed that, at 1977 prices, if effluent treatment costs were £20/tonne of dry-matter output, a 16 per cent level of incorporation would reduce feed costs for the consuming stock by 9 per cent. The researchers' proposition was to treat half of the slurry from a herd and re-feed it to the other half. This would halve the amount of slurry from the herd, slightly reduce feed costs and avoid any problems of build up of dissolved salts etc where complete recycling is used.

This work continues and the technique is yet to be proven, but in principle it would be no more difficult to engineer and operate than a swill boiler. The reassuring element is the large degree of overkill of pathogens which makes for a good safety margin under working conditions.

Another piece of research work relating to re-feeding pig manure is being carried out on a pig farm near Nottingham. The idea is to collect pig slurry; precipitate solids by adding a few per cent of lime; aerate the resulting supernatant which can then be fed back to the pigs. The liming shifts the pH to 11 which kills pathogens effectively. The work is at a very early stage and it must be recognised that results from laboratory scale work do not always materialise in full-scale application.

Other concepts to recycle nutrients in excreta include ensiling cattle manure with grass or maize, and treating excreta with alkali or mild acid to break down cellulose or lignin (fibre) to allow micro-organisms to convert some of the material to nutritionally available compounds. Very often aerobic digestion is associated at some stage with liquid treatments. At present, these are of academic interest to most farmers but as feed costs rise and anti-pollution requirements tighten such 'new fangled' ideas could well become more widely applied. Meanwhile, it is important that the necessary research work continues so that relevant practical information is available when needed.

REFERENCES

Profitable Utilisation of Livestock Manures. Booklet 2081, Ministry of Agriculture (1979).

Chapter 10

OTHER FARM WASTES

DAIRY AND YARD WASHWATER

IT IS not uncommon to find the rainwater from yards and the water used for cleaning out parlours, dairy premises and other yarded areas disposed of by drainage to a convenient ditch or stream. In many cases this has gone on for decades without any apparent real problem, and it is therefore difficult for the farm staff to accept that it is a polluted discharge.

The analysis of the liquid may be anything up to 2,000 BOD and 8,000 SS which is some way over the permitted Royal Commission standard of 20 BOD 30 SS. Under existing law such a discharge is not permitted into a water course even if the ditch is usually dry.

The long-term aim for water authorities under the COPA 1974 will be to assess and licence all discharges to watercourses. Come that happy day, what can be done about this type of discharge?

The first move is to ensure all clean roofwater is piped to a ditch or soakaway and not added to the yard water.

Estimating yard and washwater volumes

Tables 7 and 8 contain data on washwater volumes for milking premises and equipment and can be used as a basis of estimating. It is far better, however, to record accurately on a number of occasions the actual quantities produced on the farm in question. Accuracy of *actual* usage will be improved if the recordings are made unobtrusively and so avoid farm staff subconsciously economising on water leading to untypical results.

The more demanding task is to measure accurately those areas of yards, collection yards, etc which will contribute rainfall. Having done so, use may then be made of the detailed rainfall charts mentioned in Chapter 4, Table 9. It is probably prudent to plan on the once in five years' return period. From whatever source the amount of rainfall is determined, Fig. 22 can be used to determine storage volume required. Having decided the lagoon dimensions, it is easily forgotten to provide for the rainfall which falls on the lagoon itself, and so additional depth equivalent to the rain should be added.

An example

As a guide, the following hypothetical dairy herd can be considered: 100 cows cubicle-housed and slurry-scraped to above-ground steel slurry store; outside feeding area also scraped by tractor into slurry store, but due to site levels, rainfall flows away from slurry store to a gutter system also serving collecting yard. The herd is milked in a 10/10 herringbone into a bulk milk tank.

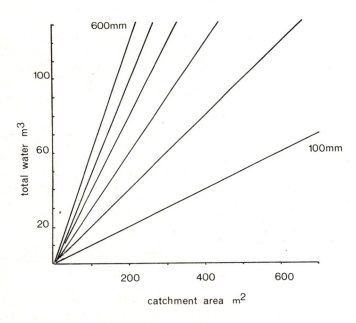

Fig. 22. Rainfall, catchment area and resulting total volume of rainwater.

Cleaning parlour (power hose)	450 l each milking
Cleaning collecting yard	350 l each milking
Cleaning units	500 l each milking
Cleaning bulk tank	220 l per day
Udder washing	200 l each milking
Other washing in milking premises	1,000 l each milking

Thus at each milking there will be about 2,720 l of washwater, of which only water from udder washing will be produced over the milking period. The remainder will be discharged almost entirely in the hour following milking, perhaps with the exception of the bulk tank which is often emptied later in the day. Therefore, the hourly rate of flow expected could be 2,720 l *minus* udder washing *minus* bulk tank washing, *i.e.*, $2,720 - 200 - 220 = 2,300 \text{ l} = 2 \cdot 3 \text{m}^3$.

To this must be added rainfall from the collecting yard, in this case 75 m^2 and from 270 m^2 of the feeding yard. Using the data from Table 9, once in 5 years 20·3 mm will occur in one hour. This would produce $20 \cdot 3 \div 1,000$ metres depth of rain over $270 + 75$ square metres of yard concrete totalling $20 \cdot 3 \div 1000 \times (270 + 75)$ m^3 of run-off, or 7·0 m^3. The calculated hourly flow of waste water amounts to 2·3 m^3 from washing, etc and 7·0 m^3 from possible rainfall, totalling 9·3 m^3. This, then, determines the minimum size of any primary sedimentation tanks or the required hourly pump performance where pump transfer is necessary. A tank with 1·5 m liquid depth, 3 m long and 2 m wide would suffice for the example quoted.

Three options

Having collected the yard water, the options are:
- to treat it to allow its re-use for yard washing and so reduce the total volume of the problem;
- to store the liquid and irrigate it to land;
- to somehow treat it to a standard high enough to allow its direct discharge to a watercourse.

(a) *Partial treatment*

With very dilute yard effluent, acceptable quality for re-use in yard washing (not milking premises) is possible by pumping off the supernatant from a large lagoon. The 'purification' which goes on is largely one of settlement of solids and, if the lagoon is large

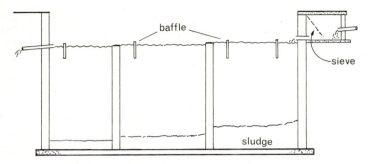

Fig. 23. Settlement tank (below ground).

enough, some reductions of BOD. It is preferable to have a sedimentation tank to receive the yard water before it reaches the lagoon. In this way the grit and heavy solids are trapped at a point convenient for their removal rather than fill up the lagoon which is much more difficult to de-sludge. Sediment tanks (fig. 23) should be sized to provide about an hour's residence time and so some idea of the flows occuring is needed before detail planning.

(b) *Store yard effluent, etc, and pump to land*

Presumably the cheapest storage will be in an earth-walled lagoon, permission for which is likely to be needed in the future from the water authority. It is unlikely that the lagoon would be of such a size (13,600 m^3 or 3 million gallons) as to come within the reservoir legislation on design and safety. Many of the aspects of lagoons have been dealt with earlier (Chapter 5) and so only two points ought to be re-emphasised, both about safety.

The lagoon should be securely fenced to keep out animals and, above all, children. Excessive depths of storage are a hazard as well as provoking problems with groundwater. It is an extra expense, but the provision of a lifebelt in a prominent position and notices warning of deep water is to be recommended if only for peace of mind.

Pumping out a lagoon needs no special mention as pto pumps and irrigation pipes and guns are commonly available. For smaller lagoons, electric pumps and small bore piping can be used to supply low level sprinklers set up in some convenient position for disposal of the liquid which will have no fertiliser value and will usually be odour free. Automatic operation is quite straightforward and is well worth considering since this may save considerably on the size of storage lagoon needed. If automatic pump-out is used, then siting of the sprinklers must be carefully chosen to ensure there is no chance of surface run-off into streams. Inevitably automatic pumping takes place soon after the rainfall which not only raises the level in the store (and so triggers pumping), but it has of course also soaked the fields. In prolonged wet periods, it may be a precaution to manually control the pump.

(c) *Complete treatment*

The problem is to reduce both the BOD and SS to what are very low levels indeed if discharge to a water course is the intention. Whilst technically this is possible, it should be recognised that it requires expensive plant, regular maintenance and a great deal of expert supervision for success. The real rub is that the standard of discharge must be achieved *constantly* and under practical farming conditions this just cannot be guaranteed. In principle the effluent is sedimented, aerated, sedimented again and perhaps aerated again before discharge. When it is realised that crude domestic sewage has a BOD around 300 mg/l (compared with yard water at 2,000 BOD) and the size, complexity and cost of the works needed to treat domestic sewage, is taken into account, the unlikelihood of achieving complete treatment on a farm becomes clear.

In the present state of knowledge, complete treatment by artificial or mechanical means of even a mild agricultural waste is a dream better left alone or at least left to those with time and money to spare.

BARRIERED DITCH

However, there is a natural means of treatment available known as the barriered ditch but this is suitable for only the mildest of effluents. The 'efficiency' of a barriered ditch is claimed to exceed 90 per cent in terms of BOD reduction. Admitting that the process is a biological one and therefore subject to fluctuations, 90 per cent might be regarded as on the optimistic side. However, even if the discharge standard is 20 BOD, the maximum BOD permissible in the influent to a barrier ditch would only be 260–250 BOD, allowing for some removal of BOD in the final length of free-running section.

Before sketching over the salient aspects of a barriered ditch, the warning must be given quite unequivocally—a barriered ditch is expensive to construct and since the water authority must agree the quality of the final discharge to the receiving water course, it would be folly to begin work on building a ditch before the agreement and written consent of the authority had been obtained. In seeking water authority consent, the opportunity should also be taken to obtain their technical advice too.

It would perhaps also be as well to list those wastes which are unsuitable for barriered ditch treatment:
seepage from livestock buildings or a manure store; poultry waste; silage liquor; dead animals or manure from animals; old engine oil; sheep dip or spray chemicals; contaminated milk or whey.

All of these materials are too strong for ditch treatment.

The Barriered Ditch System

There are three parts: a primary settlement tank; a length of ditch divided into several sections to impede flow; a final free-flowing length of ditch leading to the discharge point. In greater detail—

- The settlement tank is best sited where convenient access by vacuum tanker can be obtained to remove sludge periodically. Size should give at least one hour's residence time for the flow expected, the larger the better since de-sludging need be less frequent. A suggested design is shown in fig. 23, dimensions being determined by capacity required.
- The barrier section is intended to permit the maximum settlement of suspended solids and to allow biological breakdown of organic matter under anaerobic conditions. The residence time should be as long as possible and ninety days is a good target, the absolute minimum being sixty days.

    It is very important to prevent solids or scum from the first barrier section passing into the second and so forth as this passes on solids along the system. The solids are then broken down releasing soluble nutrients further down the ditch so increasing BOD levels.
- The free-flowing section of ditch is where aeration of the liquid can take place. The maximum length allowed by the site should be used for this section. At least 300 m should be available and allowing for the barriered section, about 400 m length is necessary in total.

OTHER FARM WASTES

Siting

It is a common misapprehension that the line of an existing ditch can be used. This is quite wrong as the flow of water would defeat the object of sedimentation in the earlier barriered section. Therefore, a new line is needed for a barriered ditch, well away from running ditches and also not in the shade of trees as they deposit an unwelcome extra organic load by way of leaf fall. Access to both sides of the ditch is highly desirable and any land drains crossing the line should be diverted.

The site needs to have enough slope (1 in 100) to enable water to flow through the barrier section and thence on down the free-flowing section to the receiving watercourse. Where such a site cannot be found near to the farm steading, then it is possible to pump the liquid from the settlement tank some distance to the head of the ditch.

Not every soil is suitable for a barriered ditch and certainly soil type will determine the batter or slope of the ditch sides. Clay soils can maintain 45° slopes and lighter soils need more gentle slopes. Close inspection of other established ditches about the farm on the same soil type will demonstrate what batters are likely to be stable. Advice from a local drainage expert would be worthwhile too.

The depth of the winter water table is important. The bottom of the ditch must be above this or else the whole ditch will operate flooded—and unsuccessfully.

The ditch profile (fig. 24) will be determined by these factors. As a guide the width of the bottom may be 1–3 m and depth and top width depending on required capacity, soil type and side batter.

Construction

It is very important that the barriers are true barriers to flow, *i.e.*, waterproof, and this can be difficult to achieve in practice.

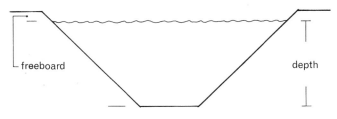

Fig. 24. Profile of barriered ditch.

- Leaving undisturbed land as the barrier between sections is a possible solution where soil type is favourable. This implies that the three to five barriered compartments will be dug out separately rather than digging a long ditch and then erecting concrete block or sleeper and polythene barriers along it. Top width of an earth barrier needs to be about 1·5 m, bottom width being determined by batter of the sides.

  To convey liquid through the barrier, a 150 mm diameter plastic pipe is installed at the designed water level. The input end should be fitted with a 'T' piece to draw from just over half depth. The outlet should be long enough so that falling liquid does not erode the downstream face of the barrier.
- A dam of sleepers is an alternative. This is not the simple job it looks. The structure must hold up a considerable weight of water and be waterproof. The edges of the dam must be located well into the bank and base, then caulked with clay as this is a frequent escape route around the ends of the barrier. The upstream face can be waterproofed by a heavyweight plastic or butyl sheet. Flow is through a baffled notch as in fig. 25.
- Concrete block wall or concrete poured in situ. Competent structural advice is needed to determine whether or not steel reinforcement is needed to resist the stresses built up in the dam wall. This type of construction is slower and more demanding of farm staff than sleeper dams. It is much more difficult to prevent leakage around constructed dams since, to provide space for working during the erection, the bank must be cut away.

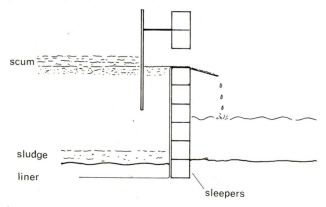

Fig. 25. Barriered ditch: timber sleeper barrier.

Replacing this excavated material adequately compacted is not easy.

Internal scum barriers positioned between the main barriers are perhaps easiest made of sawn sleepers, although suitably braced marine grade plywood would suffice. These scum boards should reach to just below half of the liquid depth.

Maintenance

A neglected barriered ditch will soon begin to fail to meet the discharge standards required by the water authority. The following suggestions will help ensure the best ditch performance:

(i) If there is a settlement tank at the beginning of the system, de-sludge it regularly. Experience will show how frequently is necessary, but every three months is a target initially. A vacuum tanker is usually the easiest, but a muck pump of the diaphragm type can be hired by the day.

(ii) The barriered sections are best cleaned out twice a year. If scum passes through from one compartment to the next, it is a sure sign that a clean out is necessary. Where total containment of flow within the ditch is intended for summertime, then de-sludging ought to be timed for the end of summer so that accumulated sludge is not flushed out by the first heavy autumn rains.

(iii) Inspect the barriers regularly for leaks.

(iv) Vegetation on the banks should be kept under control and trimmings prevented from falling into the ditch—they are an unnecessary added organic load. Growth control or other herbicides are better kept away from the ditch as they may interfere with its functioning.

(v) Bank and barrier repairs should be made promptly.

Safety

The ditch's barriered section especially will be an attraction to children and dangerous for them. An effective fence is therefore necessary which will also keep out animals. Above all, the fence must be maintained properly.

DEAD LIVESTOCK

Those casualties which cannot be disposed of to the knackers traditionally were buried in a well-limed pit. Burying in the farmyard muckheap was also practised but probably is not an acceptable method of disposal nowadays. Where animals have succumbed to disease, there is a risk of spreading pathogens.

In well-drained land, a convenient arrangement is to excavate a vertical shaft two or three metres deep and line it with 1·2 m diameter sewer pipe. The top is fitted with a secure air-tight lid and the whole construction made fly-proof. Animal carcases, afterbirth, etc may then be added as necessary and occasional inspection made to ensure the pit and contents are kept moist. A few buckets of water in summer may be necessary. After a time, a digestion of contents takes place so providing a safe and hygienic disposal facility. One such pit on a 300-cow farm has met all disposal needs for five years and still has a useful anticipated life.

Perhaps the only precautions are: to ensure the site chosen cannot give rise to pollution of underground water sources; and being a closed underground chamber, there will be a shortage of oxygen at the bottom. No one should enter the chamber without a safety line and an air supply and two assistants standing by.

Incineration of carcases is always a possibility and machines designed for the task are commercially available. However, these are mainly used by establishments engaged in animal experiments. Apart from their capital and running costs, the biggest danger lies in smell nuisance generated by the resulting plume of chimney exhaust.

An apparently successful method of dealing with dead chickens has been to add them to a large airtight tank made of plastic or fibreglass. As above, a form of anaerobic digestion takes place resulting in a smelly gruel with a concentration of bones at the bottom. The contents are then periodically removed by vacuum tanker and spread at some carefully selected remote corner of the farm. The digestion period of about three to four months appears to kill off salmonella and other pathogens.

The speed of the digestion process can probably be accelerated where it is possible to duct the exhaust ventilating air from the poultry house over and around the digester. This idea is currently being evaluated.

DRUMS AND CONTAINERS

On too many farms, used containers of whatever ilk are consigned either to the nettles or to the pit-hole used for farm rubbish. That may have been a practical solution formerly but not any longer and for two main reasons: the greatly increased potency of some of the chemicals used in modern farming, and the much more pressing need to avoid polluting our water sources upon which so much demand is exerted nowadays.

The drums and other packaging in which oils, fluids, chemicals, etc are delivered to farms can be made of metal, plastics, glass and cardboard. Some of these look an easy bet to dispose of by burning but it really is quite vital to know what the contents were and whether or not there may be danger when burning from the smoke and vapour released.

The greatest concern is over old chemical containers. There is a code of practice published by the Ministry of Agriculture, prepared in conjunction with the British Crop Protection Council, which sets out the approved methods of disposal. The Code is entitled *Disposal of Unwanted Pesticides and Containers on Farms and Holdings* and copies are freely available from MAFF. There should be a copy in every farm office because of the detailed guidance it contains.

As a generalisation, the better aim is to dispose of all containers off the farm, which means relying on the local authority or a waste disposal contractor. Local authorities vary in their attitude to providing this sort of service and many refuse. Contractors of course must charge for services but it is worthwhile to enquire about costs.

Usually some disposal will be on the farm, so some points to watch are:

- Establish a used container pound. An area clearly marked off, preferably under cover and from which children, stock and pets are excluded is needed.
- Dispose of only empty containers which have been rinsed out and never use them again.
- Glass containers can be crushed in a sack and buried at least 450 mm deep in an isolated spot away from streams, ponds or boreholes. The site should be marked and recorded. Glass can be used as hardcore under concrete.
- Metal cans should be punctured, flattened and similarly buried. Aerosol cans and vaporiser strips are best added to other domestic rubbish for disposal by the usual local authority collection. Failing this, the aerosols can be collected and delivered to the local authority whilst on business trips to the local township.
- Paper and plastic containers are tempting to burn but it is

essential to refer to the above mentioned Code of Practice to check which containers can be burnt and so avoid toxic effects.

METAL SCRAP

Over the year, it is surprising just how much scrap is generated on a farm. Many farms still rely on the 'travelling people' to call and remove scrap metal but it is not uncommon to find only the high density scrap taken, leaving the lightweight sheet, etc behind on the farm.

It is probably worthwhile establishing a dump and periodically having a 'clean-up day' around the buildings and depositing metal into a skip supplied by a scrap dealer. In this way some unforeseen cash can be generated and the efficiency of disposal more in the farmers control than relying on itinerants.

MILK AND MILK PRODUCTS

Transport dislocation may occasionally result in quantities of milk needing disposal on the farm. More normally, tainted, contaminated and mastitis infected milk may amount to 20 l or more per day on a dairy farm. There may be in addition unused colostrum during calving time. With a BOD around 120,000 mg/l it is four hundred times stronger than raw sewage and should NOT be put into ditches or other natural drainage.

There are four options for disposal:

(i) *as a feed*—average analysis of milk is 12% DM; 3·3% crude protein; 3·8% fat and 0·7% ash. It is a nutritious feed for young animals and can be added to the diet of many animals but some care is needed over rates of inclusion. This is a specialist subject and the advice of a nutrition chemist is best sought.

Again care is needed to assess any disease transmission risks where, for instance, there may be TB reactors in a herd or other symptoms of disease. Where milk must be stored for feeding then addition of formalin (commonly 40 per cent formaldehyde solution) at 1 : 250 should ensure a reasonable shelf life. Copper or galvanised tanks and pipe fittings are attacked by formalin treated milk products and so should not be used.

(ii) *mixed in with slurry*. This is probably as easy as any other way provided enormous quantities are not involved. In warm weather, there is some risk of increasing smell levels as the milk is degraded.

(iii) *sewer disposal*. It would be an offence to tip milk down the sewer and water authority permission would be needed. Their concern would be to prevent the overloading of the receiving sewage works.

(iv) *onto land*—milk has about the same N content as slurry (0·5 per cent). To avoid oxygen depletion of surface layers of soil, milk should be diluted at least 1:1 and applied evenly at no more than 22·5 m^3/ha which adds theoretically 87 kg of N/ha. This should be allowed for in any later fertiliser dressings planned for the field later in the season.

By restricting application rate, the risk is reduced of too much fat on the foliage which may restrict photosynthetic activity and alter the effectiveness of any herbicides etc applied subsequently.

Unless the health of the herd producing the milk is beyond question, it is probably a sensible precaution not to accept large quantities of surplus milk from other farms for disposal on grazing grass. Arable land is the best for disposal followed by cultivation to encourage aeration of the soil to aid breakdown of the milk BOD.

After application on grazing land, stock are better kept off for a month, less if the milk was diluted. It must be obvious that surplus milk should not be applied to steep or wet or frozen land if surface run-off into ditches is likely.

OIL

A medium-powered tractor may use 100 litres of lubricating oil a year and over half this will be drained out as used oil and contaminated with fuel, metal, acids and carbon. Being an organic material, oil will breakdown gradually if composted but the practical risk is of its escape to pollute water courses and sources.

Where a number of tractors are operated, there may be a financial incentive to save the waste oil for collection by one of the oil reclamation companies. Understandably, they tend to be more interested where sizeable quantities are available regularly, as from the garage trade.

On most farms, with a workshop, pack-house or other buildings where occasional heat is required, waste oil is burned in one of those roaring but warming vertical stoves specially designed for the job.

If oil must be disposed of on the farm, then burning is probably the best way and a small stock of oil is useful for setting off bonfires of hedge cuttings, old hay, straw or tree trimmings. Old oil is useful for brushing over bright wearing parts of ploughs, cultivators or baler knotters at the end of the season and parts of some types of sprayer pump and fertiliser distributors can be stored in oil.

SILAGE EFFLUENT

After milk, silage effluent is the strongest pollutant produced on a farm in terms of its BOD value which is usually taken as being around 65,000 mg/l, although results exceeding 80,000 BOD have been known. This puts silage effluent over two hundred times stronger than untreated domestic sewage.

When silage effluent gets into a water course, it rapidly kills off fish and most forms of aquatic life. This is because the effluent is a concentrated source of nutrients for bacteria and other micro-organisms. The arrival in the stream of so much extra food greatly increases bacterial activity which may then very rapidly use up all the oxygen in the water. Without dissolved oxygen present in water, fish suffocate.

Of the 1,293 incidents of agricultural pollution recorded by the ten water authorities in England and Wales in 1977, silage effluent accounted for 195 or 15 per cent. From the unfortunate polluter's point of view, the problem with silage effluent is that its presence in water has such a dramatic effect—dead fish and foul smells cannot be disguised. The courts will usually impose a fine which in practice is often not so very great—often under £100. The real penalty comes subsequently in that either the water authority or the local angling club will want the polluted waters restocked with fish which *is* expensive. Since the court has already apportioned blame, the polluter is identified and must pay up. To protest merely means further legal processes and costs with almost nil chance of avoiding the total bill anyway.

From all this, it is quite plain that silage effluent must at all costs be prevented from reaching watercourses or other supplies.

Quantity produced

This is related to the moisture content of the herbage ensiled as shown in fig. 26. Other factors are the type of herbage, its maturity, the severity of chopping by the forage harvester and the degree of consolidation. In the search for top quality silage, modern techniques may result in several hundred tonnes of finely chopped, immature leafy grass being delivered to the silo in a day. The

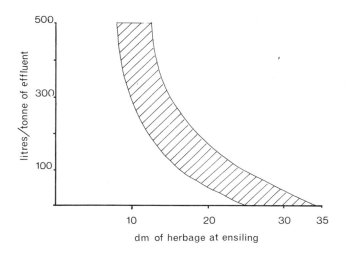

Fig. 26: Amount of effluent produced in silage making.

application of additives to the silage also results in a quicker release of effluent and an increase in the total volume too.

All these aspects need to be borne in mind when interpreting the broad curve in fig. 26. For example, herbage ensiled at 20 per cent dry matter may give between 50 and 140 litres effluent per tonne: if ensiled at 15 per cent DM, the volume could rise to 150 or 270 litres per tonne. Young leafy grass, direct cut and usually needing some additive preservative, may produce over 500 litres per tonne. Most of the effluent is produced in the first 7–10 days after ensiling, as illustrated in fig. 27.

Both of these figures clearly show the enormous benefits of avoiding low dry-matter silage-making. At around 25 per cent, little or no effluent results. By wilting the cut crop for 12 to 24 hours before picking up, the moisture content will have dropped such that very little effluent will be produced, so greatly reducing the headache of disposing of the effluent. Furthermore, there are important nutritional advantages to high dry-matter silages which should be the prime motivation to make good silage.

Usually maize silage does not produce effluent because of its high dry matter. However, where a late cool spring is followed by a cool summer, the maize crop may, as in 1978, be less mature

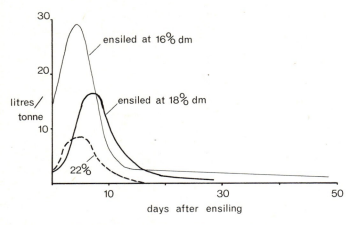

Fig. 27. Rate of silage effluent production.

than usual when harvested in late September/October and effluent in moderate quantities may result.

Collection and storage

Surface clamps are usually nowadays covered with a sheet to exclude air and so reduce storage losses of nutrients from the feed. This sheet also prevents rain leaching out effluent too. The usual arrangement is for an intercepting gulley or shallow gutter to run across the end of the concrete silo floor to collect effluent and lead it to a below-ground storage tank.

The storage tank is commonly built of concrete blocks and rendered. Because effluent is so acid, pH 4, the internal tank face should be protected with a bitumen finish or one of the specialist two-part epoxy or polymer paints or chlorinated rubber paint. The tank should have a secure lid, or if open topped, fenced absolutely securely; there should be *no* overflow fitted and the tank top should be above surrounding ground level to prevent rainwater flowing in. The best site is close to the silo to avoid long lengths of connecting pipework.

Sizing the tank is an individual problem on each farm. Where early silage cuts are taken and the weather is generally damp and if rapid harvesting rates are achieved, then a generous tankage should be provided. As a guide, something of the order of 6 to 7 m^3 per 100 tonnes of silo capacity might be considered where the effluent tank would be emptied every other day. It is shown from fig. 26 that low DM silage may produce 300 litres per tonne, and from fig. 27 that peak rate of flow occurs two to four days after ensiling at almost 30 l/t/day. Thus, per 100 tonnes capacity, 3,000 litres or 3 m^3 per day of effluent can be expected in poor silage-making conditions, and the 6 to 7 m^3 per 100 tonnes capacity looks about right. This is quite a large tank for a 800–1000 tonnes clamp and it may be acceptable to empty the tank daily to economise on construction costs.

Where weather conditions for wilting are more favourable, then a smaller tank capacity can be adequate but even then, if the silo is sited near to a watercourse, it is imperative to instal generous capacity.

Any pipework associated with effluent storage should be salt-glazed with neoprene gaskets, as cement joints are attacked. Pitch fibre and some plastic pipes are also attacked by effluent. Pipe runs should be straight and rodded out and flushed after effluent flow ceases. The storage tank should be emptied, too, and never left containing effluent. These precautions are necessary because of the corrosive nature of silage juice and the tendency for rapid growth of moulds and slime in pipes.

Poisonous Gases

This is an opportune time to mention the possibility of occasional but nonetheless acute danger from poisonous gases. Sometimes effluent is disposed of by running it into the farm's slurry handling system and it is then handled along with the rest of the manure as convenient. This is a practical solution and, given care, probably satisfactory in outside stores.

Livestock buildings are often adjacent to silos and it is temptingly convenient to dispose of the silage effluent into the slurry below the slatted floor. In these circumstances, sufficient poisonous gases can be produced to kill animals and people. There have been a number of near-fatal accidents like this.

Experimentally, it has been shown that when a small amount of silage liquor was mixed with slurry, concentrations up to 500 parts per million (ppm) of hydrogen sulphide (H_2S) were found in the atmosphere over the slurry. Thirty minutes' exposure to this level, would make a worker dizzy, prone to staggering and severe headache as the gas affects the central nervous system. Concentrations over 700 ppm could be rapidly fatal.

In similar circumstances, over 300,000 ppm of carbon dioxide (CO_2) were measured. At 100,000 ppm of CO_2, violent panting results and higher levels have a doping and toxic affect on the victim. Exposure to 250,000 ppm CO_2 for a few hours can prove fatal. The problem is there are no very obvious warning signs of gas generation since the smell of slurry masks other smells. In addition, at quite low concentrations, 2–4 ppm, the nose ceases to register the usual rotten eggs smell of H_2S. CO_2 has virtually no smell. Effervescence occurs when silage liquor and slurry mix but if the slurry is being agitated at the time of adding effluent, this warning sign too can be missed. Agitation greatly increases the release rate of the gases formed.

As a generalisation, avoid mixing silage effluent and slurry. If it is absolutely necessary then:
- Add slowly.
- When agitating prior to emptying a tank of slurry and liquor, keep personnel and stock away and increase ventilation in a building to the maximum.
- Don't enter slurry pits or tanks. If this becomes unavoidable, an air line breathing apparatus and lifeline harness must be used with at least two assistants on hand.
- Gas and dust masks are *no* protection against these gases.
- Lethal gases can accumulate elsewhere on the farm—in forage towers, grain silos. Any below-ground chamber should be treated with care.

Utilisation of Silage Effluent

As a fertiliser

Silage liquor contains plant nutrients and the amounts available are shown in Table 24, so there is some fertiliser value to be reclaimed by spreading effluent back on the land. However, because of its very high BOD value it can scorch plants, especially in hot dry weather. To avoid this, it is best to dilute the effluent 1:1 with water and spread at 20–30m^3/ha. This is easier said than done, as the slurry tanker must be half filled with water and then topped up with liquor. An overhead tank filled by mains or other water supply is quicker for filling the tanker with water than waiting for a hose to do the job. Many farms have overhead water tanks to speed up crop sprayer refilling.

Probably the best crop to receive silage effluent is the field it originally came from. Silage aftermaths are ideal in the first week

TABLE 24. Fertiliser Value of Silage Effluent

Type of silage	Dry matter %	Available nutrients (kg/m^3)		
		N	P$_2$O$_5$	K$_2$O
Grass	6	0·5–1·0	0·5	4·0
Pea haulm		2·0	0·5	4·0
Maize	8	1·0–1·5	0·5	2·5

after cutting as there is little likelihood of scorch and 25 m^3/ha of diluted effluent will supply around 13 kg N, 7 kg P$_2$O$_5$ and 50 kg K$_2$O, which will go some way to replacing nutrients removed in the harvested crop. If cows are to graze the same fields subsequently, it is worth bearing in mind the danger of applying too much potash (which may result in potash-rich foliage deficient in magnesium causing hypomagnesaemia in cows) if heavier dressings of liquor are used.

An alternative method of disposal is to use a soakaway. This should be regarded as an emergency solution. Where effluent production overwhelms the usual disposal system, the surplus can be directed to cultivated land and allowed to soakaway. Before contemplating a purpose-built soakaway, it would be a sensible precaution to consult the water authority lest the plan may endanger subterranean waters. Because of the very strong polluting power of effluent, the use of a below-ground soakaway should be very carefully investigated before construction. The very least requirement is that the water table should be well below the bottom of the soakaway, that the subsoil is permeable and there are no boreholes etc nearby.

As a feed

Silage effluent is derived not from the cell walls of plants but from the cell contents and so can be expected to be highly digestible comprising soluble nutrients, acids, proteins etc. Since 1975 work has gone on at the Agricultural Institute of Northern Ireland on the feeding of silage juice to pigs. To be useful, the effluent must be of at least four per cent dry matter and so dilution by rainfall must be avoided.

Fresh effluent of about 25 per cent crude protein was used to replace 15 per cent on a dry matter basis of the meal in a ration fed to pigs from 65 to 86 kg. Liveweight gain was unaffected and killing-out percentage, backfat thickness and hardness revealed no adverse effects. No abnormal taints were reported by a sensory evaluation panel test of the effluent fed pigs.

Effluent treated with 0·3 per cent by volume of formalin, *i.e.* 0·3 l formalin per 100 l of effluent, stored satisfactorily for 250 days in open polypropylene tanks. Moulding had begun after 300 days. Again, 15 per cent of the meal diet was supplanted by the effluent successfully and it was learned that the effluent could be introduced to the pigs quite suddenly over 24–28 hours or so without any problems and that it could be the sole source of liquid for drinking without additional water. Effluent of six per cent dry matter could be fed at 4·5 l/pig/day with a saving in meal of 0·3 kg.

Any move to use the maximum amount of silage effluent would need an appropriate system. The Irish workers used circular tanks of corrugated galvanised iron containing a butyl rubber liner. The 18 m^3 tanks were filled by pumping effluent from a small sump connected to the silo drainage system; 18 mm pipe sufficed and a 200-watt electric motor controlled by a float switch in the sump was all that was necessary. The main precaution was to ensure that pH remained below 4·5 after 1–2 weeks storage so that the formalin could prevent moulding.

On the work reported, an improvement in margin per pig of £3 occurred so that not only does the system recycle a waste, save grain and protein and reduce the risk of pollution, but there is a financial profit to be made too—a rare combination surely. However, before taking up the idea, careful consideration and costing is necessary and competent advice on nutritional and veterinary aspects should be sought.

SPRAY AND OTHER SURPLUS CHEMICALS

When using herbicides, pesticides, etc it is seldom mentioned that very often the operator will finish the job with surplus material in the spray tank—it would be a clever driver who managed to run out of spray at exactly the end of the field. The surplus can be sprayed out onto waste land, fallow or cereal stubble well away from any watercourse or ditch or neighbours' gardens or crops. The rinse-out procedure outlined by the sprayer manufacturer should be followed, again disposing rinsings onto land. Soakaways may be all right provided large volumes of persistent chemical are not involved.

It may seem trite, but before deciding a batch of pesticide, etc is surplus it is worth thinking carefully whether or not it can be used later or for another application.

If it is surplus and the container is in good condition with intact labels, the original supplier might accept its return and give some credit for it.

Where opened part-used containers are surplus, then disposal may be to land BUT only some compounds are suitable for such disposal. Fruit, bulb and potato dips as well as seed steeping chemicals present special problems. The list of which chemicals are suitable for which disposal method is contained in appendix 4 (it lists 156 chemicals) of the *Code of Practice for the Disposal of Unwanted Pesticides and Containers on Farms and Holdings* published by the Ministry of Agriculture Fisheries and Foods. Detailed advice on disposal is contained in the code which can be obtained from MAFF, Publications Branch, Tolcarne Drive, Pinner, Middlesex HA5 2DT.

Sheep dip

All sheep must be dipped twice a year against sheep scab and so all flockmasters will have spent dip for disposal. The main concern is about the persistent chemical hexachlorhexane (HCH) which is the main scab control agent, although there are other controls too.

HCH is slowly degraded in the soil and so disposal routines rely on this. Needless to say, dip should *not* be disposed of into waterways, to which should be added any place where there is risk of surface run-off into watercourses, field under-drainage systems or in the catchment area of boreholes, wells, springs or ponds. Dip should not be poured into any sewer system either, as the chemicals may well cripple the treatment works.

A soil soakaway may be used for disposal provided the construction and siting are acceptable to the water authority. The difficulty often is that most sheep areas tend to have thin soils over rock, so there is not the depth of earth to accommodate a soakaway.

The alternative is to spread the spent dip over the nearest suitable level area of land that:

(i) offers no risk of run-off or seepage to watercourse or sources, sewers, etc;
(ii) is not accessible to people and livestock, and
(iii) is not to be used for growing crops or grazing.

In practice this is a pretty tall order especially in hilly land where every scrap of level terrain is generally prized and farmed. It is worth discussing with a local forester, disposal thinly amongst his trees as a possibility. The remaining alternative is to sound out the local water authority since some have offered a tanker collection and disposal service, acknowledging the difficulties of satisfactory dip disposal in some areas and their own priority to prevent pollution of streams feeding to reservoirs.

STRAW BURNING

All that needs to be said here is that which is contained in the NFU Straw and Stubble Burning Code, which runs as follows:

The NFU Straw and Stubble Burning Code

BEFORE BURNING

1. Make a fire break at least 15 m wide by:
   Removing the straw from a 15 m strip around the perimeter of the field and ploughing a minimum of 9 furrows or cultivating this strip *thoroughly* to ensure that the stubble is properly buried.
2. Inform:
   (a) Your neighbours in order to prevent unnecessary alarm when burning starts.
   (b) Your local Fire Brigade well in advance. Know the nearest source of water and prepare directions to locate the site of burning in case of emergency.
   (c) Public owners of neighbouring land and industrial buildings, *e.g.*, British Rail (station or area manager) as well as the local Public Health Department if burning near to a residential area.
3. Remember that it is an offence to start fires within 15 m of a public highway.
4. Listen to the weather forecast and take particular care if there is likely to be an increase in wind strength or a change in direction.
5. Avoid burning in holiday areas and near busy holiday roads on Sunday and Bank Holidays.

WHEN BURNING

1. Start early in the day.
2. Burn against the direction in which the wind is blowing. If for any reason this is impracticable, burn at least 30 m at the down wind end of the field before starting to burn with the wind.
3. Have a mobile water tanker (*e.g.*, slurry tanker) or water filled crop sprayer with hose attachment on hand. Find out how you can obtain help in an emergency and have fire beaters available.

OTHER FARM WASTES

4. Put an experienced person in charge and never leave the fire unattended.
Never burn more fields than can properly be controlled.
5. The fire must be completely out before leaving the field.
Return later in order to make doubly sure.
6. If the fire seems to be getting out of control, call the Fire Brigade immediately. Meet the Fire Brigade at the roadside and show them the best way to reach the fire.
7. Remember that strong fire creates its own wind currents, particularly in still conditions. This may lead to fire spread in any direction.
8. Keep children away from the field being burned.

NEVER BURN

1. In exceptionally dry conditions.
   —In strong winds;
   —In any other circumstances when the fire may get out of control.
2. Between sunset and sunrise. Fires at night are dangerous and may alarm the public.
3. If the wind direction is likely to cause a nuisance from smoke and smuts.
4. Close to:
   —thatched buildings;
   —farm buildings, industrial buildings or plants;
   —hay or straw stacks;
   —standing crops;
   —any other concentrations of inflammable material.
5. Without adequate fire breaks to protect, in particular:
   —woodland, hedgerows or any wildlife habitat;
   —trees growing in fields;
   —electricity or telegraph poles;
   —any other public utility installations.
6. Near roads, especially motorways and busy main roads where drifting smoke could be a traffic hazard.
7. On peaty soils which may catch fire.

SPECIAL CARE IS NEEDED

1. When burning extra large fields. These should be burned in sections.
2. In the vicinity of aerodromes.
3. Near oil and gas pipeline installations.

TO OBTAIN THE MOST EFFECTIVE BURN

1. Use a spreader on the combine.
2. Combine in strips or, if cutting round the field, use a 'binder type' turn.
3. Leave the spread straw until the sap is out of the stubble (usually 7–

14 days) except when it is important to burn soon after harvest *e.g.* because of presence of wild oat seeds.
4. Keep grain trailer tracks to the minimum.

Remember that a number of county and district councils now have bye-laws to control straw burning. Check if bye-laws have been introduced in your district and if so, always burn in accordance with the bye-laws or you may be committing an offence. In the majority of cases, compliance with this code will lead to compliance with the bye-law.

Straw and stubble burning are a very important aid to arable farming. In particular they facilitate post-harvest cultivation, reduce weed infestation and help to control certain straw-borne diseases. The more effective the burn, the greater the benefits with no increase in the risks involved. Farmers, therefore, have a vested interest in ensuring that the operation is carried out efficiently, effectively and safely.

Farmers also have a special responsibility to preserve the country, its landscape and its wildlife. They must ensure that they do not cause nuisance and danger from smoke or smut to neighbouring properties, users of the highway and members of the public. Heavy claims have been made against farmers as a result of fires that have been started by sparks and from damage arising when burning has got out of control. Make sure that you are not faced with such claims—that means TAKE CARE.

Anyone starting a fire MUST observe the precautions of the safety code.

VEGETABLE WASHING WATER

The last decade has seen the growth of grading and packing of vegetables on farms for sale to supermarket organisations and a very necessary part of the preparation is the washing of the vegetables. Processing, taken to mean the removal of the vegetable skin perhaps followed by dicing, etc, is now on the increase although formerly not regarded as an agricultural operation. Processing on farm premises is not expected to become as universal as has washing. Nowadays, all the carrot crop, most of the radish, parsnip, celery and beetroot and increasing proportions of the potato crop are being washed.

The amount of water used depends on many factors including: the crop, weather and time of year (potatoes get dirtier through the winter) and especially the soil type. Washing carrots from peat soils will use twice as much water as on mineral soils. The design of the washer can be very important too: one popular and effective washer uses 4·5 m³/t/h of throughput yet a rival machine requires only 0·7 m³/t/h plus a fresh water rinse of about 0·3 m³/t/h. Table 25 summarises data from a number of sources showing the water consumption and quality of effluent resulting from farm and factory vegetable washing operations.

It is believed that the quantities used in some cases are higher than the minimum needed because where water is obtained from a watercourse, there appears to the farm staff no need for economy as the used water is returned to the system further along the flow path. This is quite usual in the East Anglian Fens. Should proposed legislation come to pass (COPA 1974) and if water authorities enforced the usual standards of effluent discharge to watercourses, there would be a number of vegetable growers with a large problem of disposal on their hands.

Minimising the problem

The basic problem is that large volumes of fairly mild (by agricultural standards) pollutant must be treated. Thoughts of

TABLE 25. Washing and Processing Vegetables—Water Usage and Waste Characteristics

Vegetable	Production period	Process	Volume (m³/t)	Effluent BOD (mg/l)	SS (mg/l)
Beetroot	Sept–Feb	Canning	4·5	4,000	1,250
Broad beans	July–Aug	Canning	6·7	300–700	—
Carrots	Aug–Mar	Washing (mineral soil)	2·7	240	4,120
		Washing (peat soil)	5·5	80	2,190
		Canning	5·6	1,400	2,000
		Dehydration	17·8	1,220	700
Parsnips	Oct–Feb	Washing	2·5	—	—
Peas	June–Aug	Washing	1·6	8,000–9,000	—
		Freezing	18·0	1,000	—
		Canning	6·7	1,100–4,000	—
Potatoes	May–Mar	Washing			
		Canning	5·6	1,400	2,000
		Lye peeling	16·5	2,570	1,000
		Dehydration	30·0	300	1,200
Spinach	April–May	Canning	32·0	300	600

Source: K. A. Smith (ADAS)

some sort of activated sludge plant like a domestic sewage works meet three immediate difficulties: many vegetable enterprises operate for only a few weeks at a time and it takes many days to weeks for the necessary biomass in a biological treatment system to establish itself: no agricultural enterprise can withstand the cost of such a treatment works; and the pollution load to be treated varies daily in volume, and in the content of soil and vegetable debris.

The most important step is to reduce the volume of water used to the minimum since this will result in smaller tanks and lagoons and a lesser volume for ultimate disposal. Waste water should be screened as soon as possible to remove vegetable debris before all soluble organic constituents are leached out into the water. Adequate sedimentation is the remaining pre-requisite of tackling vegetable waste water. Depending upon the flows involved, the local soil type and watertable, settlement tanks may be built in concrete broadly along the lines of Fig. 23 or as lagoons—but the difficulty of easily de-sludging lagoons should be remembered.

Table 26 shows how rapidly sedimentation takes place and the graph of the results suggests that most effect has been obtained after twenty minutes' residence time with a small extra gain until thirty minutes have elapsed, after which no practical reduction in suspended solids occurs. This is a useful guide in sizing a sedimentation tank when the target residence time for the flow passing through should not be less than half an hour and prefereably a little more. The reason for the over-sizing is that as the working period goes on, sediment accumulates in the tank and effectively reduces the residence time. This would be reflected in a poorer effluent quality as the day goes on.

Unless settlement tanks are sized generously, they will need de-

TABLE 26. Vegetable Washing—Settlement of Solids

Time (mins)	Carrots	Potatoes
	Suspended solids (mg/l)	
0	3,350	2,500
10	510	300
20	370	255
30	—	245
40	248	—
60	245	180
100	240	100

Fig. 28. Farm built sediment tank (capacity 3·15 m³/m width) for vegetable washing water.

sludging frequently, with dirty crops, perhaps daily. This may be by vacuum tanker or tractor-mounted slew loader fitted with a sludge bucket. The equipment available for de-sludging determines to some extent the type of settlement facility. Neither of the two pieces of equipment mentioned could tackle a large lagoon but, given care, might deal with a narrow lagoon/wide ditch.

Where concrete tanks are built, the nominal depth is about two metres and the area sized to provide the required capacity. To ensure the influent travels the maximum distance whilst in the sediment tank, two or three flow guiding boards to direct the liquid in a zig-zag fashion may be used. These boards project about 100 mm above the liquid and to half depth of the liquid, stretching alternatively from one side to the other minus a metre or so. Influent is piped or channelled into the tank through a sloping trash screen to filter off leafy materials, etc. The screen may be of wood slats or metal bars and covered with plastic mesh 4–6 mm.

A farm design of settlement tank which was simple and effective is illustrated in fig. 28. This came about becaue only a short washing season occurred on the farm and the available tool for de-sludging was a tractor front-end loader. The farmer built the tank as a simple wedge so that tractor could drive in: simple and effective.

Systems for Disposal

Having thoroughly sedimented the effluent, the disposal options include:

(i) *Discharge to the public sewer*. The water authority may not accept the type of effluent and will charge the costs of treatment at the works so that there is little advantage. It is, however, always a useful precaution to enquire about sewer disposal where it is available.

(ii) *Soakaway*. There is usually too great a volume of liquid for a soakaway, blind ditch or a barriered ditch.

(iii) *Land spreading*. This must be the easiest and most attractive on a farm where sufficient area of land is available without risk of run-off or surface water pollution. The volumes handled require an irrigation system of pump, pipes and rainguns. The modern automatic travelling irrigators can cover a hectare at one setting and are a good way of avoiding the unpleasant manual task of moving static rainguns to avoid excessive application.

This all sounds simpler than it is. The problems come in winter when the soil is wet and the danger is in surface ponding thus making for cultivation problems in the spring. If waste water is applied to growing vegetables, it is possible to transmit disease widely and rapidly.

(iv) *Treatment to reduce BOD and SS.* The reasons for contemplating this may be: to reduce BOD and so reduce the pollution load on the soil where maximum volumes are to be disposed of; to eliminate odours from a storage lagoon and when the contents are spread; and to aim for dischargeable standards, which is straightforward for BOD but the difficulties remain with SS.

The requirements are that the influent pH should be 7–8 and adequate nutrient supply and ratio *i.e.* BOD:N:P of 100:5:1 maintained which may mean the addition of inorganic nutrients. An adequate supply of dissolved oxygen to maintain 10–15 per cent saturation is also necessary and this may be provided by a floating aerator, a biological tower or a gravel bed.

Providing the conditions above are maintained, and that loading rate is not excessive, a conventional biological gravel filter can be suitable. For example, loading for a filter contained screened medium nominally 2·5–4·0 mm of about 100 gBOD/m^3/d can result in good BOD purification. Jones (see ref. p. 227) reported using such a filter to treat the mixed waste waters from the preparation of vegetables for drying. A filter 3·7 m diameter and 1·8 m deep was filled with gravel 1·9–2·5 mm size. Influent of 240 BOD applied at 0·5 m^3/m^3 of filter per day at first gave poor results but after a few days matured to give an effluent of 10 BOD after a subsequent settlement.

Filtration through sand filters is very effective but these quickly block up with organic matter. There is much hope held out for upward-flow gravel filters. In principle, influent is settled and then caused to flow gently up through a bed 150–300 mm deep of gravel in sizes 3–4 mm up to 10–15 mm grade. Flow rates are nominally 1 m/h through the bed. When the filter begins to lose efficiency due to blocking with sludge, the flow is stopped and the filter bed backwashed with a hose, the sludge being allowed to settle on the bottom of the tank, from whence it is removed by tanker or pump for disposal on land. This type of filter is currently on test at the MAFF Stockbridge House Experimental Husbandry Farm, Cawood, in Yorkshire.

Reduction of suspended solids

This is usually the real headache when attempting to clean a pollutant up to reach dischargeable standards of less than 30 mg/l. Very often recourse to chemicals is necessary to encourage flocculation of the finely divided SS. Aluminium chlorohydrate is effective, but ferric sulphate is cheaper and usually to be preferred on cost and convenience grounds. There are other formulated flocculants available commercially but their cost usually rules them out. The rate of addition of flocculant varies with the SS problem and experience soon dictates the correct dosage. Flocculant should be added to points of turbulence in the pollutant flow to ensure good mixing; a period of settlement is then needed for the solids to settle out of suspension.

Achieving dischargeable standards

To successfully manage a system of full treatment of vegetable waste water demands knowledge from the realms of microbiology, chemistry, physics, etc. The process is a biological one and is therefore influenced by changes in the strength and volume of influent and by the weather. To expect *consistently* to reach dischargeable standards is a snare and delusion in most farm circumstances. The factors are many, in considering what degree of treatment should be attempted with wastewater, and most farm situations are unique. There is, therefore, a strong case for seeking competent individual advice.

NON-FARM WASTES

There are a few such materials which are commonly found on the farm and a note about two is apposite.

Brewers' grains

These are usually wet and stored in a clamp. A moderate amount of effluent may be squeezed out as the tractor builds up the clamp. The effluent has a BOD of around 30,000 and so some care is needed over its disposal. Easiest place is in the slurry store but *not* in a barrier ditch or watercourse. The lorries carrying these grains are sometimes fitted with a catch tank to prevent effluent dribbling on the highway. Naturally enough, the driver expects to empty the tank after unloading on the farm. It is important to ensure he does not empty his tank in the farm ditch as the farmer might then be held responsible if there were consequential loss following pollution.

Swill

The collecting and feeding of swill is not only a profitable way of keeping pigs, but is also a most useful service to the community. Sometimes there are complaints of smell about the boiling process from neighbours. Most often the excessive smell occurs because of excessive boiling of the swill.

The necessary heat treatment (99°C for one hour) is usually provided by heating the swill using steam injection. When the swill is cold, a great deal of steam can be admitted and this reduces delay in getting up to temperature. However, once boiling point is reached only about ten per cent of maximum steam is necessary, since theoretically only the heat loss from the container plus a little evaporation of water has to be balanced.

With many swill boiler installations relying on manual controls, the operator is not always aware of the need to reduce steaming rate once boiling commences. If maximum steam is continued, a great deal of agitation of the swill occurs, large volumes of water vapour are produced and volatile compounds from fats, etc begin to be carried away to taint the neighbourhood.

The cure is simple and almost always effective—turn down the steam supply.

Where the swill farm is in a particularly sensitive urban area, then further smell reduction can be achieved by leading the exhaust outlet from the swill cooker into a tank of water. The pipe should discharge near the bottom of the tank so that exhaust vapours have the longest exposure to washing possible—at least 1 m depth is necessary, the deeper the better. The smell control is partly due to washing and partly due to condensation of low boiling point volatile compounds from the fat, etc in the waste food. Not all smell will be removed but a useful reduction is likely.

The washwater should be changed at three-day intervals, possibly more frequently in hot weather. The foul water will be a pollutant and is probably best disposed of on land or in the urban context into the sewers with water authority permission. The inlet pipe from the cooker should be fitted with an anti-siphon arrangement so that when the cooker cools, fouled washwater is not drawn back into the swill.

REFERENCES

JONES, E. E. Disposal of Waste Waters from the Preparation of Vegetables for Drying. *J. Society of the Chemical Industry.* **64** (1945).

Chapter 11
AGRICULTURAL ODOUR

THE PROBLEM of smells from farming operations can be likened to a ticking time-bomb under agriculture. Complaints are becoming not only more numerous but are being more effectively made. The real problem is that there is no agreed level of smell which is 'reasonable' and so the courts are in difficulty in judging fairly. A complaint made under the 1936 Public Health Act has merely to show that at the stated time and place the odour complained of was such as to interfere with the complainants enjoyment of certain rights, *e.g.*, to live without unreasonable interference from a neighbour's activities, and the court, having satisfied itself as to the accuracy of facts presented, will almost automatically find for the prosecution.

The foregoing is an over-simplified description, of course, since before reaching the courts there is usually dialogue between farmer and complainant often extending over a long period of time. Usually there is more than one complainant, in which case the local authority environmental health officer may be the (unfortunate) go-between twixt polluter and offended parties.

If a complaint is made about water pollution, a crucial piece of evidence is the measurement made of any alleged pollutants. In absolute contrast, when odour is the complaint, no measurements are usually available and so the polluter is in danger of being required by the courts eventually to reduce the odour emission until no complaints are likely. This is a nebulous concept and in practice, once someone gets a dose of foul smell sufficient to take action about it, he tends to become sensitised not only to smell itself, but also to activities he can see which he associates with foul smell. The best example is slurry spreading, whereby complaints have been received by farmers who were on the occasion in question spreading water (in fact a telling tactic to de-value a complainant's evidence subsequently in court!).

In practice, therefore, the polluter may have to reduce odour to almost nil level to guard against repeat complaints. Like the human species, animals themselves smell as does their excreta and it is not reasonable to expect economic livestock production to go on in the absence of smell. So here is the conundrum facing the courts, the polluter and indeed society at large: how shall 'reasonableness' be accommodated and defined in an odour complaint situation, especially in view of the highly subjective nature of what is judged to be a foul smell?

SMELL—WHAT IS IT?

Like the other four senses, smell is registered by the brain via electrical pulses sent to it from the relevant detection areas of the body. Smell is detected in the nasal cavity which reaches from the nose about 75 mm back to the top of the throat. The smell detectors are minute filaments about $0 \cdot 1$ μm (micrometre) diameter and 100 μm long (*i.e.*, one tenth of one millimetre) and are situated in the olfactory cleft which is high up at the back of the nasal cavity.

This explains why slow shallow breathing may be used to avoid an unpleasant odour as the air by-passes the olfactory sensors. Conversely, to detect a smell or odour better one usually takes a good sniff, thus delivering air fully into contact with the odour-sensing nerves. The action of smell is not well understood but it is likely that odoriferous compounds dissolve in the mucous membranes of the olfactory cleft and the resulting solution of ions influences the electrical potential of the sensors, so generating the electrical changes registered in the brain.

There are three concepts associated with smell: strength, description or character and pleasantness.

Strength. This relates to the concentration of the smell and is measured in terms of its 'threshold dilution value', which is assessed by a group of people known as a smell panel. Comprising six or eight representatives chosen to avoid odour-insensitive individuals, the panel is presented under prescribed laboratory conditions with odour samples of known dilution alternated with blanks of clean air. Each member independently indicates whether or not he can detect a smell; when half the panel cannot detect the smell, then that particular dilution is rated as the 'threshold dilution value'. It may be measured as a few hundreds or several millions depending on the strength of the odour.

The technique, by which odour samples are taken on site and transported long distances to the smell panel, demands skill and experience. It is interesting to find marked differences in methods used by continental and British researchers, and it is to be hoped that sufficient funds are made available for UK work to catch up with Europe on agricultural odours.

Character of an odour is its description. An onion smell is different from curry, bad drains or scent. Distinguishing radically different smells as these is simple, but a large element of personal judgment occurs in deciding between two different but similar smells.

A particular smell may be made up of many different chemical compounds. The Warren Spring Laboratory (WSL) of the Department of Industry has done a deal of research and evaluation work on industrial odour and has identified more than 150 compounds thought to contribute to odour. Commonly occurring compounds include aldehydes, ketones, skatole, indole, mercaptans and so on. These chemical compounds are formed once manure goes anaerobic. Theoretically, foul odours are prevented by keeping the manure supplied with oxygen during its storage. In practice, the liquid manure from housed livestock will always be anaerobic as the manure usually accumulates in depth within the livestock housing and is not removed for some days.

Most of the odour compounds are reckoned to be in the soluble BOD fraction (determined by centrifuging a sample to precipitate solids) of the manure. If this BOD requirement is met by adding oxygen, then smell is controlled.

Pleasantness, or more often unpleasantness, of an odour is hugely influenced by the individual perceiving the odour. To coin a phrase 'one man's scent is another man's peculiar smell' and it is doubtful if they could ever agree. Strength, character and pleasantness can themselves interact to change one's attitude to a smell. A faint whiff of bad drains may be identified without the recipient reacting unduly. If the smell persisted or increased in strength, undoubtedly a complaint would result. Furthermore, someone dwelling downwind of an overloaded sewage works would perhaps not notice the earlier mentioned drain smell until higher strengths were reached yet he might well notice the faintest tinge of odour from a FYM heap.

Thus different smells affect people in different ways and will be rated differently on a nuisance basis. First experience of a smell may be rated differently from subsequent exposures. The conditions under which an odour is perceived—cold, hot, windy, dry or wet weather—will also change nuisance rating. Even from this superficial sketch of odour, it becomes clear just how knotty a problem faces the farmer receiving a complaint about odour.

SOURCES AND CONTROL OF FARM ODOUR

Most complaints about smell concern livestock farming and the handling, storage and spreading of manure, etc. Cows and poultry are involved, but by far the greatest concern is aroused over pig farms, the aroma from which is foul, clinging and able to travel far. Badly-made silage can be a problem, as can cabbage fields in autumn, but both of these are transitory and do not usually last long.

(a) About Buildings

In passing, it ought to be mentioned that the sale of houses or of building land for houses close to farm buildings merits *very* careful consideration. It is quite possible to result in odour complaints from the new owners.

Manure heaps, especially if wet, generate smell and flies. Keeping the heap as small and contained as possible is all that can be done, given that the site has been chosen carefully. Whilst it may be convenient to site a heap in or around buildings, if necessary the heap may have to be established elsewhere on the farm. Heaps are usually not a great source of trouble although large-scale composting for mushrooms can bring odour problems. It is important to spray the heap against flies as problems with them may turn into odour complaints if domestic housing is nearby.

Manure stores may be a problem. A compound is not often a cause for complaint probably because it is undisturbed. It cannot practically be controlled for smell and this may be a reason for choosing another system. Lagoons of pig muck can give off a deal of odour in warm weather about which little can be done. Choice of site is important and attempts can be made to empty the slurry in cold weather, but this is not always possible as land may not be available.

Tank storage of slurry may give rise to smell during agitation, especially during hot weather. The use of jetting kits when silos are very full may give rise to manure aerosol which can carry a long way. Operation of these or any other agitation device should be restricted to favourable weather conditions.

Where a complainant lives very close to the tank the options are to offer to use a masking scent or agent, cover the tank or move the tank, in ascending order of difficulty. Aerosol dispensers to release masking agents are not expensive (£100) and once a scent is found acceptable to the complainant, the dispensers can be switched on whenever wind direction suggests.

The difficulty with complaints about smell around the farm buildings is that having spent money on alleviating the initially alleged source of odour, there is no guarantee of freedom from complaint again since there are so many possible sources of smell about the buildings.

The exhaust ventilating air from livestock houses can be a potent source of smell in warm weather, especially from broilers in the last two or three weeks of production. There have been reports of improvement by reversing airflow direction so that exhaust is at low level instead of at the ridge. This may be all right where one or two sheds are concerned, but with larger numbers the basic problem remains of very large volumes of odorous air emanating from the site as a whole. Competent engineering advice should be sought where airflow changes are contemplated, as ventilation volumes and patterns may be unintentionally changed.

On the Continent, limited experience has been gained of a biological filter to control exhaust smells. All exhaust air is ducted to a large box in which is housed a medium, usually plastic polynettes (like ladies' hair curlers), over which water is trickled and up which passes the odorous air. Initially inoculated with activated sludge from a sewage works, a biomass builds up on the medium and which digests odour compounds dissolved in the water. The sizing of the medium is aimed to give a contact time between odour and biomass of 0·6, preferably 1·0 seconds. As will be deduced, because of the large volumes of ventilating air necessary in summer, this requires a very large filter. So far as is known, such a filter has not been operated in the UK.

Again on the Continent, Van Geelan and Jongebreur of IMAG at Wageningen have reported work on biological air washer/filters to control odour from pig houses. Commercially available equipment based on IMAG designs has been marketed in Germany and no doubt could be imported into the UK. However, an approximate price of £2,000 for 100-pig places will doubtless slow down purchasers. The general arrangement is as shown in fig. 29 and a unit is positioned beneath each extractor fan. Liquid recirculation rate is 2 l/s and fresh mains water automatically maintains liquid level, replacing liquid lost by evaporation and the 300 litres of liquid removed from the system daily to control sludge content. Correctly sized and maintained, these units are very effective, no odour being detectable alongside a piggery so fitted.

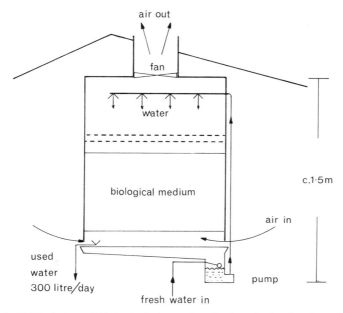

Fig. 29. IMAG-designed biological air washer for installation in piggeries. One unit beneath each exhaust ventilating fan.

Alternative possibilities are to pass the exhaust air through a soil, compost or other biological filter from buried perforated pipes. Work has been done on soil filters at Warren Spring Laboratory and design parameters are expected but no full-scale unit has been operated as yet.

Problems associated with such filters are (i) the collection and ducting of the odorous air to the filter, (ii) the need for very low resistance to airflow lest existing ventilation fans become inadequate and (iii) the need to keep the filter damp not wet. Indications are that a very large filter area may often be needed and which must, of course, be fenced off from animals. There is a need for research and development on biological odour filters to establish their economic and practical feasability under UK conditions.

(b) From Field Operations

Odour complaints mainly erupt as a result of field operations, more frequently from applying slurry by tanker or raingun than

TABLE 27. The Influence of Weather Conditions and Odour Dispersal

DAYTIME

Wind	Nil	Sunshine Weak	Bright
Light	2	3	4
Moderate	3	4	5
Strong	4	5	5

NIGHT

Wind	Clear	Cloud cover Broken	Overcast
Light	1	1	2
Moderate	2	2	3
Strong	3	3	4

Source: L. G. Bird, ADAS. Crown copyright.

Key: 1 = least dispersion conditions
5 = most dispersion conditions

from solid muck spreading. Control methods may be technical or tactical by nature, of which the latter is much cheaper and easier.

Tactical control covers common-sense precautions like not spreading too close to boundaries, nor on weekends, public holidays or days preceding them if warm weather is expected when neighbours will be enjoying their gardens or have windows open. Overdosing with manure will result in smells remaining much longer to annoy the neighbourhood.

Above all, the weather can be a useful ally if weather forecast information is studied, especially concerning wind direction and temperatures. ADAS Meteorologist L. G. Bird produced a simplified guide (Table 27) to those weather conditions giving good odour dispersion and so favourable for manure spreading. Odours disperse slowest when the night is clear and the wind is light, which generally occurs when barometric pressure is high. A 10 km/h wind will mix odour at ground level with upper air for only a hundred or so metres. In contrast, on bright days with fresh winds rapid odour dilution occurs. A 50 km/h wind will mix the air to a height exceeding a kilometre. Furthermore, shower clouds are accompanied by strong rising air currents which can reach five kilometres height and rapidly dilute odours in so doing. Thus a strong wind in the direction away from sensitive neighbours is always helpful, and spreading is best done during a bright day—perhaps a statement of the obvious—but the merit of Table 27 lies in the attempt to categorise the in-between weather conditions. An additional benefit is to focus attention on the need to study the weather and its likely dispositon for the two or three days following intended slurry spreading.

Technical control includes some form of treatment of the manure to prevent its smelling on application to land or afterwards.

AERATION

Removal of odour whilst the manure is in its storage container is one possibility and this is usually done by aerating the slurry with an electrically-powered mechanical aerator (Plates 30–32). There have been a few installations where the aeration is carried out in the circulating slurry channel below the slatted floor using a specially-designed aerator. A rotary impeller is driven by an electric motor via an extension shaft direct from the motor, and air is sucked down the shaft by the action of the impeller submerged in the slurry (fig. 30).

The aerator is run continuously and will then cope with the manure from about 200 pigs.

With electricity at 3p/kWh, power costs per pig would amount

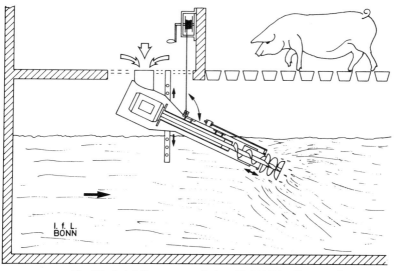

Fig. 30. Axial-flow aerator/mixer (2·2 kW by Puremex).

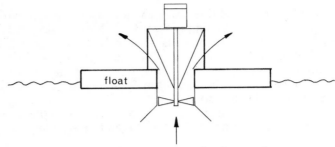

Fig. 31. Upward throw floating aerator.

to 95p. The aerator costs £880 or say £1,000 installed and written off over five years at 20 per cent interest, allowing only 5 per cent for repairs would bring the cost per pig to £1·57p. Another possible drawback is the need to dilute the slurry to maintain adequate speed of flow around the continuous channel to prevent sludging. However, the manure does not smell when spread on the land and the atmosphere in the pig house is much improved.

Observations in Germany have reported pig performance being retarded by gases from slurry.

The more usual place to aerate slurry is in the storage tank or lagoon. A floating aerator of 10–15 kW costs £2,000 and more plus

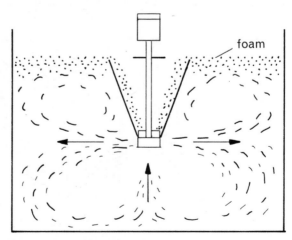

Fig. 32. Down-draught floating aerator: thickness of surface foam is controlled.

costs of supplying electricity to the site. What simpler than to throw on an aerator, tether it, switch on and leave it to get on with aeration. Sure enough, if oxygen is forced into the manure, the smell will be reduced eventually but this is a long way away from a carefully and correctly designed system.

Scientifically speaking, there must be a correct amount of oxygen needed to control odour *i.e.* the minimum amount in the interests of least costs and avoiding unnecessary wasted energy. Investigations on this aspect have been carried out at the West of Scotland Agricultural College using laboratory scale digesters. This has led to suggested rates of oxygen, and attempts are under way to monitor practical installations to compare results.

At the practical end, questions remaining are about the oxygenation efficiency of aerators. Most of the manufacturer's figures are based on performance in water. In particular, it would be useful to know what volume of slurry can be mixed adequately by an aerator. The power needed to mix the slurry will be much greater than that needed for actual aeration. How should the aerator be operated—continuously or intermittently? How does the farmer know when he has aerated enough? What type of automatic control is offered? At present, UK suppliers of aerators cannot give this information, although it is known that one or two are concerned at the lack of information and are making efforts to remedy the situation.

It is not uncommon to find manufacturers suggesting that the aerator is switched on 'a few days' before the manure is spread. This may give an odour reduction during spreading but at the price of intense odour levels around the store for the first three to six days after switching on. This sort of experience can not only be a disappointment to the farmer, but may in fact bring renewed or even fresh odour complaints.

There is a pressing need for development work on *systems* of aeration odour control. The subject is complex and both residence time, quality of the effluent and especially the amount of dilution present will influence the required rate of aeration for smell control. As a rough rule of thumb, with an aerator sized 20 W/m^3 German sources suggest it takes about 10 days' aeration per 100 m^3 of slurry to control odour and about a two weeks' start-up period in addition. (Dutch work suggests a continuously run aerator sized about 6 W per pig place for odour control.) Professor Wolfermann and his workers at Fachhochschule des Landes, Bingen, West Germany reported only a quarter of his smell panel subjects could perceive any odour at 25 m from slurry spread after continuous aerobic treatment and no odour at all at 50 m. In contrast untreated pig slurry was rated offensive at 500 m distance by over a third of the panel. There are undoubted advantages to separating manure before aerating as the free-flowing liquor can be much more effectively mixed. It cannot be too strongly stated that competent technical advice is sought before investing in an aeration system so that the individual farm problem can be diagnosed and a specification drawn up of aerator requirements. This is a more businesslike approach than buying a large aerator, incurring excessive electricity consumption and never realising that odour control can be just as effective and much cheaper.

Field experience with aerators has not been altogether uneventful. Motor failures, shaft breakage, impeller loss and vibration fatigue of the machine have all been reported. A practical safeguard is that if any parts are replaced, they should be weighed and compared with the old part. After all, the whole device must float—and there is at least one story of a failed continental motor being replaced by a UK 'nearest equivalent' which when released promptly sank!

ANAEROBIC CONTROL

It is well known that sewage works subject the domestic waste to an anaerobic treatment. Their motives are twofold: to control smell and to produce a stabilised sludge. Undoubtedly, anaerobically digested animal manure has very much less smell than raw slurry. This is because of the very large BOD reduction (80 per cent) of the process which eliminates all soluble BOD and reduces much of the non-soluble 'tougher' BOD. Currently it is common for this to be reported as smell *control* in the vein of elimination. This is quite misleading. Digested manure does have some smell, reminiscent of creosote and not unpleasant, but caution is needed when spreading because there is no guarantee that such smell may not be a cause for complaint.

Lagoon stores of digested liquor seem to give off very little odour and are reported by Scottish researchers to be stable for many months. In practice, the degree of odour reduction offered by anaerobic digestion is very noticeable and clearly a considerable asset when investing in a biogas plant which has been dealt with more fully earlier. The smell control element is unquantifiable financially and must depend on the farmer concerned and his

TABLE 28. Odour Control—Masking and Other Agents

Agrigest	Badgett Cook Biochemicals Ltd 27 Emperors Gate London SW7 4HS Tel: 01-373 9431	1 lb £5·50 Dilution rate: 1 lb in 1 gal water for slurry
Alamask MX	Lautier Aromatiques Ltd Power Road Chiswick London W4 5PJ Tel: 01-995 0555	5 litre tin £17·75 25 litre drum £83 Dilution rate: 2 pints in 10 gals per 10,000 gal pit 4 fl. oz in 2 gals per 1,000 gal tanker load
Deodorant M (range of scents)	Northern Aromatics Ltd Dumer Lane Radcliffe Manchester M26 9GF Tel: 061-796 7209	1 kg £1·50–2·60
Hyzyme	Hydrachem Ltd Billingshurst Sussex Tel: 040-381 3598	1 kg £11 Dilution rate: 2 kg treats 1,000 gals for 17 weeks
Maskomal	Antec International Chilterns International Estate Sudbury Suffolk Tel: 078-73 77305	1 kg £17·91 Dilution rate: 6 ozs in 1,000 gals (repeat every 10 days)
Pit Stop	Salsbury Laboratories Cremyll Road Reading Tel: 599233	1 gal £18·10 Dilution rate: 1% dilution for 7 days continuously, then once per week
Sani-Lisier	The Mews Minting House Horncastle Lincs LN9 5RX Tel: 065-86 220	1 gal £14 Dilution rate: 1 part: 4,000
Sectalam (odour control & insecticide)	Lautier aromatiques Ltd (as above for Alamask)	1 kg £4·50 Dilution rate: 4% by volume

interests and priorities. In discussion about biogas production, it is usually the combination of energy production *and* smell control which accelerates an enquirer's interest.

CHEMICAL CONTROL

This description is used loosely here to include control agents not strictly chemical in their action.

In passing, it ought to be mentioned that pure chemicals can be added to slurry to control its smell. They work usually by adding oxygen, *e.g.*, adding hydrogen peroxide, ozone or ammonium persulphate. The problem is one of chemical cost and they are not economic.

Adding ordinary lime at 5 per cent will raise the pH to 11 or 12, and assist sludge settlement leaving both sludge and supernatant with very little smell. Such a system has not yet been operated at large scale as an odour control. The likely difficulty would be in handling rather a lot of wet running sludge—and in paying for the lime.

A variety of 'agents' are on the market for adding to slurry. Descriptions such as biological extract, dried fortified biomass and so on are common. Claims are made to control odour and sometimes to reduce the viscosity of the manure, making it easier to pump. Just a tinge of doubt crosses the mind sometimes when looking at the handful of material which, after mixing in a few litres of water and thrown into a slurry store, will over a period of weeks, it is claimed, radically transform a smell problem.

None of these materials examined at laboratory scale has proved to be all it might be. Of course, it would be quite wrong to here brand all such materials as useless. Perhaps the moral is to carefully read any claims and at best try a small lot first of all. It must in fairness be reported that one pig farmer, troubled with smell both from a lagoon and when spreading on autumn stubbles especially, was convinced he had reduced both problems by treating all slurry cellars every three weeks from April with a biological agent. Each treatment cost £50 for around 7,000 pig places.

Masking agents

The use of masking agents has long been practised by water authorities around overloaded and smelly sewage works. A variety of scents can be supplied by manufacturers and examples are given in Table 28.

In principle they work by substituting a hopefully less disagreeable smell for the problem odour. As they are volatile compounds they must be emitted continuously whilst the masking effect is needed, and so for large sites they can be expensive in material costs.

These agents invariably produce the smell they claim but that is not the end of the problem in practice. First of all, not every nearby resident will like the particular scent and may, in fact, complain about the mask odour itself. Another problem is that being chemical compounds, they may react with or combine with compounds in the odour to produce new compounds, some of which may just make the original smell appear a mild titillation. In short, it is possible for masks to produce undesired effects. Furthermore, there is the possibility of partial separation out of the scent at some distance downwind, possibly causing annoyance.

The general reaction of water authority personnel experienced in the use of masking agents seems to be pessimism in that they are no permanent solution to an odour complaint situation. However, they are a useful short-term psychological weapon against an odour complaint because their use is proof that the polluter is taking positive steps and this is an important tactic. It is a matter for individual judgment whether or not to offer to use the masking agents immediately since if they are successful, the polluter may be forced to use them thereafter. This may be more expensive long term than other options available.

Masking agents can be dispersed by aerosol generators on top of buildings or, in some cases, by mixing a small dose into the empty slurry tanker before filling. The scent is then dispelled with the slurry. Rose, pine and other scents are available.

Interestingly, in some continental countries masking agents are frowned upon as they themselves are regarded as chemical pollution of the environment.

FIELD EQUIPMENT FOR CONTROL

- Very good control of smell can be obtained by spreading slurry carefully and immediately ploughing or cultivating it into the soil. This is seldom used in this country because of finding the correct weather, crop and soil conditions and it is time consum-

ing. However, it should not be forgotten as a possible last resort for odd occasions.
- Slurry tankers should have the lowest possible height of spread (Plate 16) to minimise drift. There are a few on the market and a working demonstration is the surest way to be satisfied.
- Direct injection equipment for slurry was described earlier with all its limitations. However, the control of odour is impressive in the field.
- Tankers fitted with rear-mounted dribble bars (Plate 29) are effective in generating very little smell.
- A simple but very effective odour control attachment to a tanker is shown in Plate 45. Slurry is delivered into the box section distributor bar and falls between the two plastic sheets or skirts to the ground. No aerosol is generated and the slurry is laid onto the ground in a strip. The advantages are that the tanker work rate is not impeded, no extra power is needed and the attachment is cheap (£250—1979 price). Any farmer faced with odour complaints of aerosol from field spreading ought to consider at least a demonstration of the device.

ON RECEIVING AN ODOUR COMPLAINT

It must be recognised that a genuine serious complaint can be pressed through the courts and eventually win, given persistence and application by the complainant. It is counter-productive to try to brush off and ignore a complaint as usually the victim resides in the area and is unlikely to move. Of course, not all complainants are balanced and mature in presenting their odour complaint; many are in an excited nervous state by the time they have screwed up their annoyance to do something about the foul smell endured. The most costly mistake of all is to lose personal control as this converts a difficult situation into a personal feud and reason flees.

The complainant must have a courteous hearing and it is important to immediatley express concern about his problem. By discussion it must be established:
- Who is complaining, one or how many and where resident.
- How the smell is described.
- When the smell occurs: date/s, time/s.
- The effects of the smell: headaches, sickness, taints of food or clothing.

After the complainant has gone, the likely cause of the odour should be determined and studied. Are particular operations the cause and how important is wind direction?

The local NFU representative should be informed of serious odour complaints to keep the NFU in the picture on behalf of the industry. The cause should then be rectified as far as possible. Advice should be sought widely as to the various options open. If there is likely to be delay in effecting improvements, it is very important to keep complainants fully informed.

Getting off an odour complaint hook is not easy and much can be done to help by good psychological tactics:

—good public relations are never wasted;

—if operations likely to generate odour are vital, choose least sensitive times and inform complainants (note through letter-box);

—farm transport carrying muck or slurry through the village must leave no spillages and the vehicles should be clean and driven with consideration;

—there is nothing wrong in seizing the initiative in the right circumstances and inviting complainants and other residents to tour the farm (tidied up beforehand) as a whole, stressing those aspects showing concern for the environment as well as those of food production. Some small refreshment—not too lavish—to complete the tour provides an opportunity for people to talk which is beneficial—we, all of us, complain most effectively against unknown people but not nearly so forcefully against those we have met a few times.

INDEX

Accident, 40
Acts (of law), 31
ADAS, 20, 50, 234
Advantages of separation, 159
Advantages of slurry, 73
Aerators, 165–6, 176–8, 235–7
Aerobic, 24, 165, 195, 235
Aerosol cans, 207
Aerosol dispensers, 231
Agitation of slurry store, 64, 65
Agricultural inspectors, 39
Agricultural (Safeguarding of Workplaces) Regulations, 39
Agriculture Act, 187
Agrigest, 239
Alamask, 239
Anaerobic, 24, 165, 168, 191, 238
Article 4 Directions, 35
Augers, 123
Automatic scraping, 114, 117
Availability of nutrients, 57, 58, 59

Bales, 45
Barriered ditch, 201
Bedding materials, 43
—, usage, 45
Best practical means, 34
Biogas production, 168, 170, 172, 178–9, 191
Birmingham University, 169
Boar, 42
BOD_5 definition, 24
—, figures, 41
—, test, 23
Brakes, 127
Brewers' grains, 225
British Crop Protection Council, 207
Broiler, 42, 45
—, litter, 45, 185

Building Regulations (1976), 35
Bulk milk tanks, 47
Bull, 42
Burning straw, 218–20
Bye-laws, 33

Cadmium, 24
Calculating application rate of FYM, 147
Calf, 42
Capacity of slurry stores, 100
Carbon dioxide, 24, 214
Carcases, 206
Centrifuge, 155
Charges for sewer disposal, 32
Chemicals, spray, 216
Clean Rivers (Estuaries and Tidal Waters) Act (1960), 31
Cleaning equipment, 47
—, water, 47
—, yards, 48
C:N ratio, 169
COD, 24
Code of Good Agricultural Practice, 37
Code of Practice, (disposal of pesticide containers), 217
Composting, 25, 167
—, temperatures, 151
Compounds, 77
—, with earth floor, 79
Concrete pipe stores, 97
—, stores, 83, 97
—, tanks, 90
—, —, sprayed construction, 96
Controlled waste, 36
COPA, 25, 31, 36–9, 51, 75, 221
Copper, 25
Costs, biogas production, 174–81
—, handling manure, 130–2

245

Countryside Act (1968), 35
Court, High, 33
—, lower, 37
—, magistrates, 39

Dairy washwater, 197
Dead livestock, 205
Deep pit, 96
Definition of farm waste, 15
Denitrification, 25
Deodorant M, 239
Deposit of Poisonous Waste Act (1972), 33
Diastase barley, 188
Digestion, 25
Discharge (of effluent), 31, 38, 225
—, pattern (of slurry), 128, 142
Disposal pit, 206
Dribble bar, 153, 162, 194
Drums, 206
Dry matter, 25, 122, 126, 152, 157, 171, 186, 211, 215
Drying manure, 185, 193
Ducks, 42, 53
Dumping, 36

Earth-banked compound, 77–83
Eberhardt Silopresse, 189
Economic life (of equipment), 131
Effluent, silage, 210, 212
Electricity generation, 172
Elevator, 123, 159
Ensiling cattle manure, 195
—, (grass), 211
—, poultry litter, 187
Eutrophication, 26
Excreta production, 21, 41, 42
Explosive mixture, 173

Facultative bacteria, 26
Farm Storage Ltd, 62, 63, 84
Farm waste, definition, 15
FCGS, 35

Feeding litter, 189
—, silage effluent, 215
Fencing, 90
Fertiliser and Feedingstuffs Regulations (1973), 187
—, prices, 59
—, value (silage effluent), 215
—, —, (separated liquor), 154
FHDS, 35
Fiat Totem, 172
Financial value of manures, 59
Flame trap, 173
Floating aerator, 176–8, 236
Flocculation, 225
Flushing, 110
Formalin, 216
Four-wheel-drive loader, 113
FYM, 26, 43–4, 52
—, financial value, 61
—, handling, 112, 138
—, losses, 54
—, midden, 71
—, nutrients, 53
—, storage, 68

Gases, poisonous, 213
Generation of electricity, 172
Glass waste, 207
Goose, 42

Handling FYM, 138
—, slurry and manure, 104
Health and Safety at Work Act (1974), 39, 173
Health and Safety Executive, 39
Heat recovery, 166
Heifer, 42
Herb sizes, 16, 17
Herbicides, 216
High Court, 33
Horse, 42
Hose, 48
Hydrogen sulphide, 213
Hyzyme, 239

INDEX

IMAG (Holland), 193, 232
Injection of slurry, 133, 143, 144
Irrigation of slurry, 136, 145
Irrigator, 136–7, 161

Kale, 183

Lagoon, 62, 74, 110, 199, 238
—, agitator, 125
Legislation, 30
Lime, 195, 240
Loading ramp, 86
Local Government Act (1972), 33
Losses in store, 54, 55
Lough Neagh, 57

Magistrates, 33
Magnesium, 26
Maize, 195, 211
Management of stores, 102
Manure analysis, 53
—, drying, 185
—, pollution strength, 18
—, production, 21, 41
—, utilisation, 182
Masking agent, 231, 239
Maskomal, 239
Medicines Act, 187
Metal scrap, 208
Meteorological Office, 50
Metrication, 22, 23
Midden, 52, 54, 71
Milk waste, 208
Minister of Agriculture, 37, 38
Ministry of Agriculture, Fisheries and Food, 20, 35, 50, 207, 217
Ministry of Agriculture, Fisheries and Food census, 21
Ministry of Agriculture, Fisheries and Food Technical Bulletin No. 35, 72
Mink, 42
Mixing, mechanical, 101
Molasses, 188

Newcastle-upon-Tyne University, 195
Newspaper, 44
Newsprint, 44
NFU, 32, 243
—, Straw burning code, 218–20
NIAE, 26, 52, 111, 154, 156
NIRD, 153, 154
Nitrate, 26
Nitrification, 27
Nitrite, 27
Nitrogen, 27
—, loss, 55, 56
Noise, 38
Noise Abatement Act (1960), 39
Nomograph, 106
Nuisance, 33, 34, 38
Nutrients, availability, 57, 58, 59
—, for plants, 53
—, of manures, 53, 183
NWC, 32

Odour, 16, 128, 228–43
—, complaints, 242
—, sources, 231
Oil consumption, 186
—, waste, 209
Organic irrigation, 136
Overflow channels, 105–9

Paper waste, 207
Parlour washwater, 199
Pathogens, 68, 153, 185, 187
Pattern of discharge, 128
Peat moss, 44
Pesticides, 216–17
pH, 27, 169, 188, 195, 216, 224
Phosphate loss, 56
Phosphorus, 27
Pig farm sizes, 20
Pig manure nutrients, 53
Pigs, excreta production, 42
Pit Stop, 239
Planning permission, 34, 35

Plant nutrients, 53
Plastic waste, 207
Pneumatic stirring, 101
Poisonous gases, 213
Potassium 27
—, loss, 56
Poultry housing, 95
—, manure, as fertiliser, 183
—, —, dry matter, 70
—, —, ensiling, 181
—, —, nutrients, 53
—, —, value, 185
Poultry Research Centre, 185
Power requirements, 129
Price of fertiliser, 59
Prosecutions for farm waste pollution, 19, 32
Protein Processing Order, 190
Public Health Act (1936), 33, 228
Public Health Act (1961), 32
Public Health (Drainage of Trade Premises) Act (1937), 32
Public Health (Recurring Nuisances) Act (1969), 34
Pump, centrifugal, 120
—, piston, 189
—, pneumatic, 124
—, positive displacement, 120
—, slurry, 115, 119

Rabbit, 42
Rainwater, 49, 198
Ramp, 86
RCEP, 20, 66
Reco Ltd, 63
Re-feeding manure, 184, 190
Relevant waters, 37, 38
Rivers (Prevention of Pollution) Acts (1951 and 1961), 31
Roofwater, 49
Rough terrain forklift truck, 103, 116, 139
Royal Commission Standard, 27
Rowett Research Institute, 168

Safety at Work Act (1974), 39, 139, 173
Safety fencing, 79, 90
Salmon and Freshwater Fisheries Act (1975), 32
Salmonella, 190
Sample size, 56
Sand, 44
—, filters, 224
Sani-Lisier, 239
Sawdust, 44
Scab control, 217
Scrap metal, 208
Scraping, automatic, 114, 117
Sectalam, 239
Secretary of State, 31, 34, 35, 38
Sedimentation tank, 199, 223
Seepage, 51, 54, 72, 78, 217
Separated fibre, 150
—, liquor, 137, 150, 154, 164
Separation, advantage, 159
Separator, 123, 149, 155, 159, 161
Settlement, 200, 222
Sewer disposal, 32, 217, 223
Sheep dip, 217
Silage effluent, 210, 212
—, —, feeding, 215
Silage making, 184, 211
Size of herds, 16, 17
—, —, pig herds, 20
Skid steer loader, 113, 116
Slew loader, 112
Slurry, advantages, 73
Slurry channels, 92–4
—, —, 'V' shaped, 109
Slurry graph, 105
—, handling, 104
—, irrigation, 136
—, losses, 55
—, pump, 115, 119
—, storage, 66, 73
—, tankers 126, 142, 194
Smell, 16, 52, 128, 210, 229–30
—, complaint, 242
—, control, 241–2
Sodium, 28

INDEX

Soil injection, 133, 143, 144
Solid retention time, 28, 170, 171
Sow, 42
Spray chemicals, 216
Spreaders, 140–7
Spreading FYM, 140
SRT, 28, 170, 171
Stone trap, 128
Storage of FYM, 68
—, of slurry, 73
Strainer box, 87, 149
Straw, 44
—, bale compound, 85
—, bales, 45
—, burning, 218
Swill, 41, 226

Tanker costs, 132
Tankers, slurry, 126, 142, 194
Terrington EHF (MAFF), 85
Time of application of manures, 59
Total solids, 28
Total suspended solids, 28
Totem, 172, 191–2
Town and Country Planning Act (1971), 34
Town and Country Planning Act (General Development Order, 1977), 34
Tractor costs, 131, 175
—, scraper, 117
Trade effluent, 31
Trailers, 147
Trawscoed EHF(MAFF), 190
Trespassers, 40
Turkey, 42
Tyres, 127

Underground seepage, 51
University of Birmingham, 167
University of Newcastle-upon-Tyne, 195
Utilisation of biogas, 180
—, —, manures, 182
—, —, silage effluent, 214

Vacuum tanker, 126
Value of manure, 185
—, of silage effluent, 215
Vaporiser strips, 207
Vegetable washwater, 220–5
Volatile fatty acids, 28
Volatile solids, 29
Volumes of waste, 41

Wageningen (IMAG), 232
Warren Spring Laboratory, 230
Washwater, 197
Waste Disposal Authorities, 36
Water, 46, 48
—, cleaning, 47
—, table, 51
—, usage, 221
Water Resources Act (1963), 31
Weather, 234
Wheels, 127
Whey, 53
Wood shavings, 44
Work rate, 129, 135, 140

Yardwater, 197

Zinc, 29